# Cognitive Technologies and the Pragmatics of Cognition

## *Benjamins Current Topics*

Special issues of established journals tend to circulate within the orbit of the subscribers of those journals. For the Benjamins Current Topics series a number of special issues have been selected containing salient topics of research with the aim to widen the readership and to give this interesting material an additional lease of life in book format.

**Volume 12**

Cognitive Technologies and the Pragmatics of Cognition
Edited by Itiel E. Dror

These materials were previously published in *Pragmatics & Cognition* 13:3 (2005)

# Cognitive Technologies and the Pragmatics of Cognition

*Edited by*

Itiel E. Dror
University of Southampton

John Benjamins Publishing Company
Amsterdam / Philadelphia

∞™ The paper used in this publication meets the minimum requirements of American National Standard for Information Sciences – Permanence of Paper for Printed Library Materials, ANSI Z39.48-1984.

**Library of Congress Cataloging-in-Publication Data**

Cognitive technologies and the pragmatics of cognition / edited by Itiel E. Dror.
p. cm. (Benjamins Current Topics, ISSN 1874-0081 ; v. 12)
Includes bibliographical references and index.
"Previously published in Pragmatics & Cognition 13:3 (2005)."
1. Cognition. 2. Cognitive science. 3. Artificial intelligence. I. Dror, Itiel E.
BF311.C553455 2007
153--dc22 2007029147
ISBN 978 90 272 2242 8 (Hb; alk. paper)



John Benjamins Publishing Co. · P.O. Box 36224 · 1020 ME Amsterdam · The Netherlands
John Benjamins North America · P.O. Box 27519 · Philadelphia PA 19118-0519 · USA

*I dedicate this book to my four L's and my one and only S.*

# Table of contents

# About the Authors

**Itiel E. Dror** is a Senior Lecturer in Cognitive Sciences in the School of Psychology at the University of Southampton, UK. He holds a number of graduate degrees, including a Ph.D. in cognitive psychology from Harvard (USA). He specializes in the fields of human cognition & behaviour, training & skill acquisition, technology & cognition, expertise, and biometric identification. Dr. Dror has worked in universities around the world and conducted research and consultancy to numerous organizations, including the UK Passport Services, the USA Air Force, the Japanese Advance Science Project, the European Aerospace Research & Development Agency, the Israeli Aerospace Industry, the BBC, and many commercial companies. Dr. Dror holds a UK Government ESRC-EPSRC grant on *merging technology and cognition*. He has published widely in both basic science and applied domains.

**Graham Pike** is a senior lecturer in psychology at The Open University's International Centre for Comparative Criminological Research. His research interests are in forensic psychology and applied cognition, particularly eyewitness identification and face perception. As well as the EPSRC funded work reported here, he recently completed two U.K. Home Office projects that included an investigation of using computer technology, mainly video parades and facial composite systems, to make the most of identification evidence.

**Nicola Brace** is a senior lecturer in psychology at The Open University's International Centre for Comparative Criminological Research. Her research areas include eyewitness memory and testimony and the development of face perception skills in young children. She has conducted Home Office funded research looking at how to improve the construction and presentation of facial composite images as well as field research examining the use of video identification parades.

**Jim Turner** is a postdoctoral research fellow working with The Open University's International Centre for Comparative Criminological Research. He recently completed a PhD at The University of Westminster that sought to use knowledge gained from perceptual research to improve the construction of facial composites using the E-FIT system. His work on using minimal face stimuli has since been incorporated into the development of the latest version of E-FIT.

**Sally Kynan** is a research fellow working with the Applied Psychology Research Group at The Open University and is also a member of the International Centre for Comparative Criminological Research. Her research interests include forensic and educational psychology, particularly developing and analysing interviewing techniques for use with vulnerable witnesses, and combines quantitative and qualitative approaches.

The authors are part of The International Centre for Comparative Criminological Research at The Open University. Recently they have been involved in two U.K. Home Office projects to investigate using computer technology to make the most of identification evidence and to study the visual identification of suspects. Their background is in applied cognitive psychology, particularly with regards to eyewitness evidence.

**Juan C. González** (Ph.D C.R.E.A., École Polytechnique, Paris 1998) is Professor of Philosophy and Cognitive Science at the State University of Morelos, Cuernavaca (Mexico), since 1999. He is interested in combining empirical and conceptual approaches to study (visual) perception and cognition in general. His research in this perspective includes: space representation, categorization, concept-formation, perceptual consciousness, sensory substitution, qualia, abnormal states of perception. Other fields of philosophical interest: ecological ethics and social theory.

**Paul Bach-y-Rita**, M.D., is Professor of Orthopedics and Rehabilitation and Biomedical Engineering at the University of Wisconsin, Madison, and was Professor of Visual Sciences and of Human Physiology at the University of the Pacific. He has been developing sensory substitution and brain-machine interfaces for over 40 years and has numerous publications, books, patents, Federal research grants, and national and international awards. He also specializes in brain plasticity and the theory of brain reorganization.

**Steven J. Haase** (Ph.D. University of Wisconsin, Madison) was an Assistant Professor of Psychology at Gordon College (1994–1999) and a Researcher at the University of Wisconsin, Madison helping design experiments with a team of engineers working on sensory substitution (1999–2002). Since 2002, he has been an Assistant Professor of Psychology at Shippensburg University in Shippensburg, PA. His interests include perception, human-machine interfaces, theoretical modeling of psychological processes, attention and human information processing, and consciousness.

**Stevan Harnad**, born in Hungary, did his undergraduate work at McGill and his doctorate at Princeton and is currently Canada Research Chair in Cognitive Science at University of Quebec/ Montreal and adjunct Professor at Southampton University, UK. His research is on categorisation, communication and cognition. Founder and Editor of *Behavioral and Brain Sciences*, *Psycoloquy* and *CogPrints Archive*, he is Past President of the Society for Philosophy and Psychology, Corresponding Member of the Hungarian Academy of Science, and author and contributor to over 150 publications.

**Willem Haselager** is Assistant Professor of Artificial Intelligence / Cognitive Science at the Nijmegen Institute for Cognition and Information (NICI), Radboud University. He is a regular visiting professor at the Philosophy Department of the Universidade Estadual Paulista (UNESP), in Marília, SP, Brazil. He holds master degrees in philosophy and psychology and a Ph.D. in theoretical psychology. He is particularly interested in the integration of empirical work (i.e., psychological experiments, computational modeling, and robotics) with philosophical issues regarding knowledge and intelligent behavior. He analyzed the debate between proponents of classical cognitive science and connectionism on the nature of representation, in relation to the inability of computational models to deal with the frame problem (interpreted as related to abduction and common sense knowledge and reasoning) and examined the consequences of that debate for the status of folk psychology. More recently he has extended his research by investigating the embodied embeddedness of cognition (EEC) in relation to dynamical systems theory (DST) and the theory of self-organization. Of late, he has started to explore issues in conscious will and autonomy.

**Fred Phillips** is Professor of Marketing, Entrepreneurship and Research Methods at Maastricht School of Management. He has been Head of the Department of Management in Science & Technology at Oregon Graduate Institute, and Research Director of the $IC^2$ Institute, the University of Texas at Austin's think-tank for the technology-business-society interface. He is author of *Market-Oriented Technology Management* (2001), *The Conscious Manager: Zen for Decision Makers* (2003), and *Social Culture and High Tech Economic Development: The Technopolis Columns* (2006).

**Maria Eunice Quilici Gonzalez**, B.A. in Physics (1977, UNESP, São Paulo), M.A. in Epistemology (1984, UNICAMP, Campinas), and PhD in Cognitive Science (1989, Essex), is a Professor at the Department of Philosophy at the University of the State of São Paulo (UNESP) at Marilia, where she is the director of Graduate Studies in Cognitive Science and the Philosophy of Mind. She is President of the Brazilian Society for Cognitive Science.

**Niall Griffith** received a degree in Archaeology and Anthropology from the University of Cambridge in 1972. He investigated economic prehistory before taking an MSc in Intelligent Knowledge Based systems. He was awarded a PhD in 1994 by Exeter University for research on self-organizing neural network models of musical structure. Since 1996 he has taught Computer Science at the University of Limerick. His research interests are computational musicology and cognition.

**Meurig Beynon** is a Reader in Computer Science at the University of Warwick. He received his PhD in Mathematics from King's College London in 1973 for research that underlies a categorical equivalence now known as the Baker-Beynon duality. He founded the Empirical Modelling Research Group in the 1980s and has subsequently published over 80 papers relating to this innovative modelling approach and its applications in fields as diverse as business, engineering and educational technology.

# Gold mines and land mines in cognitive technology

Itiel E. Dror
University of Southampton

Technology has long played an important role in human activity. However, with technological advances we are witnessing major changes in the role technology plays. These changes are especially revolutionary in two senses: First, new technologies are playing greater than ever roles in human *cognitive* activities. These activities include: 1. New levels of cognitive interactions between people. These interactions, both quantitatively and qualitatively, are at an intensity and scale that allow new forms of cognition to emerge, such as distributed cognition. 2. Technologies that cognitize with us, thus playing an active part in our cognitive processes and constituting themselves as inherent components in human cognition. 3. These new technologies do not only cognitize with us, but they also cognitize for us. In this sense they go beyond supplementing human cognition; rather than playing a facilitating role they actually take over and replace certain aspects in human cognition altogether.

Whether these technologies give rise to new forms of cognition, such as distributed cognition, or they cognitize with us and for us, these technologies mark a fundamental change in the role they play in human activities. Such technologies are best termed *cognitive technologies* (Dascal and Dror 2005).

The second sense in which these technologies revolutionize their role is that they are actively affecting and changing human cognition itself. In the past when they were predominantly a tool to aid humans, they had a minimal role in shaping cognition. They only played an instrumental role in executing the product of human cognition. Now, with increasing emergence and use of cognitive technologies, they are more integrated in the cognitive processes themselves. As such, they play an active and constituting part in human cognition. Since human cognitive processes are adaptive, dynamic, and pragmatic, they do not work in isolation from cognitive technologies. These new technologies affect and shape cognition.

As new cognitive technologies emerge and with their wider integration in human activity, they influence and change the very way we think, learn, store information, communicate with one another, and a host of other cognitive processes, thus changing the nature of cognition and human activity.

This new emerging field of cognitive technology is of great interest and importance. Its implications are all encompassing; they raise academic and scientific questions, as well as practical issues of how best to develop and integrate these technologies in the workplace and at home. They also introduce social, moral, and philosophical issues. It is time to investigate and consider the broad issues surrounding cognitive technologies. Technological innovations are very fast and the rapid changes they introduce are followed by legal, social, and other slower responding systems which try to consider and adapt to the technological impacts. Cognitive technologies, as I will try to illustrate, offer a great potential across many domains. However, their power and intrinsic influence on human cognition can be detrimental and harmful. Thus we need to understand and carefully consider the *gold mines and land mines in cognitive technologies*, as I explicate below.

To consider cognitive technologies, I will focus my examination of its impact mainly on two broad and fundamental domains: The first is data exploration and investigation, and the second is learning and training. Data exploration and investigation, from initial design of the methodology for collecting the data and the actual data collection, to its exploration, analysis and interpretation have all been profoundly affected by cognitive technologies. The gold mines of these technologies are that they offer great opportunities for data explorations and investigations that have never existed before. For example, in psychological experiments we can relatively easily design complex methodologies that involve experimental design to collect response time data from participants. In the past the apparatus for such experiments would require months if not years of work, but nowadays this can be achieved in a matter of days if not hours.

The data collection itself has also been affected by new technologies; nowadays using multiple computers or web based studies, hundreds, if not thousands of participants can contribute to data collection within a few hours. Even in domains that do not rely on human data, new technologies enable the collection of huge amounts of data with great efficiency. A variety of data mining technologies allow efficient exploration of vast amounts of data in very little time. In the past a great deal of effort and time was needed to collect and explore such amounts of data. Once the data has been collected and initially explored, its further analysis and interpretation is relatively trivial. Statistical packages and other software enable us to analyse and visualize data to uncover interesting patterns in a matter of minutes, if not seconds.

These examples illustrate the great power and abilities that new technologies offer; a gold mine, no doubt. However, they also introduce some potential land mines that we need to consider. Too often such technologies are embraced without fully considering (and taking countermeasures to) the problems they introduce. For example, the ease of collecting data and its analysis have reduced the investment in planning and thinking. With cognitive technologies it is so easy to carry out these activities that rather than thinking carefully how best and most appropriately to do things it is more straight forward to just adopt a 'trial and error' approach than to consider things in depth. Using this approach, for example, when you design a study rather than investing thought whether (for illustrative purposes) to expose a stimuli for 100 or 150 milliseconds, you are more prone to use one of them and 'see how it goes' because you know it is very easy to modify the exposure time. Similarly, when you analyse the data, because you are not computing the statistics by hand, you can run a variety of models and use different statistical approaches with great ease. This leads many times to not carefully considering which is the best approach, but just to try one, and if it fails, then to try another. The problems with such impacts of cognitive technology are not limited to possible distortions in the correct and scientific procedures and results (such as an increase in false positive statistical significance findings as a result of multiple testing), but has far reaching implications to the level and depth of thought put into these data investigations.

Such land mines introduced along with the gold mines offered by cognitive technologies are not limited to data investigations in the scientific domain, they are equally applicable to other domains. For example, moving from the laboratory scientific inquiry to the 'real world', we can see these implications in the forensic world of fingerprint identification. Although fingerprint identification has been around and used in courts for over a hundred years, it has been revolutionized in the past few years with the introduction of new technologies. These technologies have affected all aspects of fingerprint identification, from using scanners rather than ink to collect fingerprints, to their digitization, and the introduction of mobile devices that can do these and other functions. But most interesting and revolutionary is the introduction of the Automated Fingerprint Identification System (AFIS). These technologies enable us to take a partial or distorted fingerprint left at a crime scene and compare it against a very large set of fingerprints stored on a database. In a matter of seconds AFIS will provide the closest matching prints for a human expert to consider. AFIS offers great power, and indeed many crimes have been solved because of these new technologies, including old unresolved cold cases.

However, as we have seen in the domain of scientific inquiry, such gold mines introduce land mines. In the forensic fingerprint identification domain, the gold mines AFIS has produced have also brought about dangerous land mines in the form of erroneous identification. With the introduction of very large databases and the ability to search them via AFIS, there is now a high likelihood of finding very similar 'look alike' fingerprints by pure coincidence (Dror, Péron, Hind, and Charlton 2005). Thus, the criteria threshold for concluding an identification needs to be adjusted to the use of such powerful technologies. The erroneous Mayfield case illustrates the practical and real land mines that are introduced with these technologies. Using AFIS Mr. Mayfield was selected as a suspect in the Madrid bombing. Three fingerprint experts at the FBI examined the fingerprints of Mr. Mayfield and they unanimously and independently misidentified him as the Madrid bomber (Stacey 2004). The important point here is that the erroneous identification of Mr. Mayfield was in part because of the powerful technology of AFIS. This technology enables us to search very large databases, and thus will result in finding very similar fingerprints by pure coincidence. When such similarity exists, it is much more likely to make erroneous identification (not only in fingerprint, but in any other pattern recognition task, such as aircraft identification, see Ashworth and Dror 2000).

The last domain that I want to use to illustrate gold mines and land mines in cognitive technology is learning and training. Technology Enhanced Learning (TEL) has been used to facilitate and improve one of the cornerstones of cognition and human activities: Acquiring, storing, and using new knowledge. TEL has been taking an increasing role in almost all learning environments. It is used in a variety of informal and formal educational environments, as well as in many commercial, industrial, and governmental settings. Since these cognitive technologies are having a growing use and impact in the area of learning and training, it is important to consider some of the gold mines along with the land mines they introduce. These will further illustrate the general issues associated with cognitive technologies.

First, in general, for learning to be successful it must conform to the architecture of the mind. For example, this means training must take into account constraints on information processing capacity. Information during learning need not be reduced to fit the limits of the cognitive system, rather the information must be conveyed in ways in which the system can easily acquire and store it. This can be accomplished by using the correct mental representations and engaging the cognitive system on its own terms. Doing so will not only enable quick and efficient acquisition, but the knowledge gained will be better remembered and will have an impact on behaviour. Using TEL offers great opportunities to build efficient and effective learning programs, but the powers that TEL provides may also overwhelm the human cognitive system. Thus, they bring to the forefront the need to make cognitive technology fit and work well with the architecture of cognition (Dror 2005).

Second, when we consider specific technologies (and their usage) we need to examine what they offer as well as what they may limit. This applies to a variety of TEL in which we need to understand how the use of electronic boards and visualization tools, e-learning, synchronic vs. a-synchronic remote learning, blackboard, simulation and gaming, interactive videos and virtual realities, and other specific TEL environments affect learning and the learner. Lets take an example of a very basic and widely used tool: PowerPoint. An increasing number of learning and training presentations are provided via PowerPoint. This TEL specific tool offers a gold mine in terms of presenting information in a succinct and clear fashion. It enables us to present multi-media and complex information in an easy manner that simplifies learning. However, the use of PowerPoint has also had a detrimental affect on learning. This tool has been used many times in a very limited and expected format, resulting in boring and ineffective learning. It is not the tool itself, but the way it is used. This is a fundamental point across cognitive technologies: they offer great opportunities, but also have vulnerabilities. These gold mines and land mines are highly dependent on how we utilise these technologies, rather than on the technologies per se.

Third, and finally, TEL needs to be considered and understood in light of learning objectives: not only the acquisition of information, but also the ability to retain and use it. Learning, in all its stages, depends highly on the learners paying attention and being engaged. Learning technologies offer real opportunities in this regard. Beyond specific TEL tools, such as simulation, gaming, and interactive videos, which are designed for this purpose, all TEL enable us to promote a great deal of active learning. For example, providing control to the learners helps to achieve active and motivated learners, and when they are involved, participating, engaged, and interacting with the material, then learning is maximised. It is maximised because it activates and correctly taps into the cognitive mechanisms of learning, such as attention, depth of processing, and other cognitive elements of learning. TEL enables us to shift from merely exposing the learners to the material, to transforming the learning environment.

In terms of control, the learners can be given control over the presentation format of the material. Because learners have different experiences, cognitive styles, etc., they may have preferences for the way the material is delivered (for example, visual vs. auditory, text vs. diagrams, etc.). Giving them control over the format of presentation not only gives them control but also optimises and tailors the learning to the individual learner. At a more basic level, learners can control the pace of learning (e.g., when to move on to the next item/page, and whether to repeat a section before moving on to the next). Thus, this illustrates that TEL can help to establish active and motivated learners, and bring about engagement, involvement, participation, and interaction. These are all critical ingredients for achieving effective and efficient learning.

However, as with other cognitive technologies, TEL can have detrimental affects. It can hamper learning by utilising its powers to provide too much to the learners, and thus end up making them passive. For example, memory is probably one of the most important dimensions in learning because learning most often is aimed at conveying knowledge to the learners so they retain and remember it. TEL can hinder memory by its very nature and merit. One of the appealing elements of technology is its ability to provide information in a very effective way; many times by taking the burden off the learners. However, if not done properly, reducing the effort and work involved in learning is not necessarily good (Bjork and Linn 2006). It may promote 'spoon feeding' the material, which makes the learners more passive and decreases their depth of processing, leading to reduction in retention and memory of the learned material. The use of TEL does not only affect the efficiency of how we acquire and retain information, but it is changing how we learn and what learning is all about.

I have used data exploration and investigation and Technology Enhanced Learning to illustrate cognitive technologies and to exemplify the gold mines and land mines they introduce. These opportunities and pitfalls are --of course-- not limited to these two domains that I have used for illustrative purposes. Mobile phones are highly used technologies that have transformed how we communicate with one another, the language we use, how we access and store information, and so forth. Like the other cognitive technologies I have discussed, this device offers new and great opportunities, but also can have a variety of detrimental affects. Cognitive technologies are growing, both in terms of new technologies emerging and also in terms of their wide usage in a variety of human activities. It is thus important to consider their full impact. What we need to understand is that cognitive technologies are no longer just aids in helping humans achieve their goals, but that they are becoming so engrained into the cognitive process that they affect it and who we are.

## References

Ashworth, A.R.S. and Dror, I. E. 2000. "Object Identification as a Function of Discriminability and Learning Presentations: The Effect of Stimulus Similarity and Canonical Frame Alignment on Aircraft Identification". *Journal of Experimental Psychology: Applied,* 6: 148–157.

Bjork, R. A. and Linn, M. C. 2006. "The Science of Learning and the Learning of Science: Introducing Desirable Difficulties". *American Psychological Society Observer* 19: 3.

Dascal, M. and Dror, I. E. 2005. "The impact of cognitive technologies". *Pragmatics and Cognition* 13: 451–457.

Dror, I. E. 2005. "Experts and technology: Do's and Don'ts". *Biometric Technology Today* 13: 7–9.

Dror, I.E., Péron, A., Hind, S., and Charlton, D. 2005. "When emotions get the better of us: The effect of contextual top-down processing on matching fingerprints". *Applied Cognitive Psychology* 19: 799–809.

Stacey, R. B. 2004. "Report on the erroneous fingerprint individualization in the Madrid train bombing case". *Journal of Forensic Identification* 54: 706–718.

# Making faces with computers

## Witness cognition and technology*

Graham Pike, Nicola Brace, Jim Turner and Sally Kynan
The Open University, United Kingdom

Knowledge concerning the cognition involved in perceiving and remembering faces has informed the design of at least two generations of facial compositing technology. These systems allow a witness to work with a computer (and a police operator) in order to construct an image of a perpetrator. Research conducted with systems currently in use has suggested that basing the construction process on the witness recalling and verbally describing the face can be problematic. To overcome these problems and make better use of witness cognition, the latest systems use a combination of Principal Component Analysis (PCA) facial synthesis and an array-based interface. The present paper describes a preliminary study conducted to determine whether the use of an array-based interface really does make appropriate use of witness cognition and what issues need to be considered in the design of emerging compositing technology.

## 1. Introduction

Despite the recent advances made in physical and photographic identification, the eyewitness continues to play a central part in police investigations. That witnesses tend to be somewhat less than reliable has become a phenomenon well documented in research (for example see Cutler and Penrod 1995) and, as a result, juries are warned against placing too much store in witness testimony in both UK and US courts. Perhaps the most important information that an eyewitness can supply is that relating to the identification of the perpetrator, but unfortunately this form of evidence is just as prone to error, if not more so, than more general forms of information about the crime.

Of key importance are the cognitive processes involved in encoding, storing and retrieving the face of the perpetrator. There is little, if anything, that can be done to improve the encoding and storing stages of this process, beyond only using witnesses who got a 'good look' at the time of the event; so it is the retrieval stage that researchers, practitioners and technology development have focused on in an attempt to improve the accuracy of eyewitness identification evidence. A considerable proportion of this research and development has concentrated on the construction and conduct of identification parades. Although of great evidential value, identification parades involve fairly minimal interaction with the witness, often limited to standardised instructions and the witness' decision as to which, if any, of the people in the parade is the perpetrator.

In contrast to the single decision asked for at an identification parade, the construction of a facial composite (an image, typically constructed by recombining features from several faces, that attempts to capture the likeness of the person in question) of the perpetrator requires a great deal of interaction between the witness, the police operator and the system being used to aid construction. It is therefore vital that the interaction between witness cognition and technology be based on information that the witness can readily and accurately provide, and avoid cognitively difficult or error prone tasks.

It is possible to produce an image of the perpetrator by employing a sketch artist and this technique has been found to produce accurate results (Laughery and Fowler 1980), although it does require considerable skill and training. To overcome this requirement, systems such as Photo-FIT and Identikit were developed in the 1960s and 1970s and required witnesses to search through albums of individual facial features so that these could be assembled into a face-image. Research generally found that these early systems produced poor likenesses (e.g., Christie and Ellis 1981; Ellis, Davies and Shepherd 1978). Two explanations for the inaccuracy of the images produced were suggested: that the systems simply did not contain a database of features sufficient to cope with variability in appearance; and that the system was asking the witness to perform a task that was cognitively difficult. Although the former explanation undoubtedly accounts for some of the inaccuracy, the latter found great resonance with theories of face perception that suggested that the face was not simply encoded as a set of features but rather that recognition was also dependent on more holistic information, including the configuration of the features (see Rakover 2002 for a review). In addition, some research had found that features were easier to recognise when presented as part of a face, i.e., within a face context (e.g., Tanaka and Farah 1993). The poor results found with systems such as Photo-FIT may have, at least in part, been due to the fact that the witness had to search through individual features which were not presented within a face context: a task that did not match the essentially holistic nature of the cognitive processes involved in face perception.

Later systems, such as E-FIT, CD-FIT and more recently PROfit, took careful note of both applied and theoretical face perception research and were designed to make better use of witness cognition. To this end features were only presented within a face context and tools were included to help make alterations to the configuration of features as well as to the individual features themselves. Attempts to evaluate these computerised systems have often produced inconclusive (and sometimes very complex) patterns of results (for example see Davies, van der Willik, and Morrison 2000). It should be noted that very often laboratory based research excludes many of the factors likely to lead to an accurate likeness being produced in the field, such as trained operators, lengthy interviews and construction times, and the use of artistic enhancement through image manipulation software (Gibling and Bennett 1994). However, although this criticism undoubtedly prevents the forensic utility of any particular system from being estimated accurately, the fact that participants in the experiments find it difficult to produce an accurate image is still very interesting for what it reveals about the interaction of cognition and technology.

Looking at the generally poor rates of identification found from composites it is possible to conclude that witness memory is simply too inaccurate to produce usable images. However, as well as a simple comparison across systems, some researchers have looked in more depth at how well composite systems make use of witness cognition (e.g., Brace, Pike, Allen, and Kemp, in press) and have found that although more recent computerised systems such as E-FIT do interact more appropriately with the cognitive processes of the witness, they still involve tasks that are fundamentally difficult to achieve with any accuracy given the limitations of human cognition.

The composite construction process employed in systems such as E-FIT requires the witness first to recall the target face and verbally describe it to the operator. The operator then enters this description into the system by selecting appropriate feature descriptors, which are used by the system to rank order the exemplars for each feature and to produce an initial image comprised of the feature exemplars best matching the description provided by the witness. The witness is then shown this image, attempts to determine what is wrong with it and directs changes to the image so as to better represent the target face. This is achieved by the operator using the system to replace individual feature exemplars with others from the database, altering the relative size and position of the features and using image manipulation software to add finer touches to the face.

Examination of the above process shows it to be based on several tasks, each of which people generally find very difficult to perform with any accuracy. Firstly, the witness must recall the face of the perpetrator. Although research has found the act of identifying an unfamiliar face (such as at an identification parade) to be problematic (e.g., see Leippe and Wells 1995; Levi and Jungman 1995; Wells and Sealau 1995), recognition at least appears to be the cognitive task that human face perception processes have fundamentally evolved to perform. Recalling a face involves retrieving the memory of that face and then somehow consciously bringing it to mind and is a far more complicated and difficult process to perform (Shepherd and Ellis 1996). Once the face has been recalled, the witness must then verbally describe it to the operator. Research which has examined the descriptions provided by witnesses has generally found them to be of very poor quality, often missing out many features and containing little information about configuration (see, for example, Buckhout 1974; Pozzulo and Warren 2003). Importantly, neither recall skills nor vocabulary seem to lend themselves to describing faces in detail, leading to sparse descriptions (Fahsing, Ask, and Granhag 2004).

To improve facial compositing technology so that it interacts more appropriately with witness cognition it is therefore important to address three factors: the use of a database comprised of individual features; the reliance on verbal descriptions of the face; and the reliance on facial recall. Recently several new facial composite systems, notably EVOfit and EigenFIT in the UK, have been developed with exactly these three factors in mind. Rather than use feature databases, both systems make use of a statistical technique known as Principal Component Analysis (PCA), which works by analysing the image properties of the whole face and thus capturing information that is intrinsically holistic, just as human cognition appears to be (see, for example, Hancock, Burton, and Bruce 1996). PCA is applied to a training set of face images (usually several hundred to several thousand in size) to produce a set of eigenfaces (essentially the images corresponding to each eigenvalue resulting from the PCA). Although these eigenfaces in themselves do not represent any useful abstraction, they can be combined using different weighting mechanisms to form any face within the 'face-space' described by the original learning set. Thus by combining the eigenfaces it is possible to synthesise any face, as long as the learning set was suitably representative of the target population. The exact nature of this mathematical procedure is not of concern here; it is enough to note that the database is derived from entire faces, rather than individual features, and that construction therefore proceeds in a more holistic manner.

As no witness could describe a face from memory according to its constituent eigenvalues, an interface radically different from that used by previous systems needed to be developed (see also O'Toole and Thompson 1993). To avoid asking the witness to verbally describe the target face at any point in the procedure, the initial interfaces designed (for both EVOfit and EigenFIT) require the witness merely to select one facial image representing the closest match to the target from an array of several images (each array usually containing either 9 or 16 faces). The system then generates a new array of faces containing the previously selected face and with the other faces in the array resembling the chosen face to varying degrees and in varying ways. With each choice, the system gradually reduces the amount of variation in the array so that the faces should become more and more like the target. In fact, both EVOfit and EigenFIT make use of genetic algorithms to assist in narrowing down the variation (see Gibson, Solomon, and Pallares-Bejarano 2003 and Hancock 2000, for more detailed descriptions). As all the witness needs to do is select the closest match to the target in each array, and to tell the operator when the system has produced a good likeness of the perpetrator, there is never a need to involve verbal descriptions.

The third factor described above, that of avoiding recall, is also helped by the new interface. However, although it has become common to refer to array-based (or parallel) interfaces as being based on the cognition of recognition rather than recall, the situation is not actually quite so clear-cut. In fact the task would only be one purely of recognition if the witness looked at the first array produced and told the operator that one of the faces was an excellent likeness of the perpetrator. The chances of a good likeness being present in the first array are statistically extremely remote. Instead, the witness must look at each face in the array in turn, compare them and decide which one of them is the best likeness. When presented with the variation necessarily inherent in the initial array, this task might be akin to a pure recognition decision as the best match is likely to be easily evident. However, once the system begins to narrow the amount of variation between faces in the array, the witness' task becomes much harder and they undoubtedly have to look at each face systematically and make comparisons both between faces in the array and between each face and their memory of the target face. Thus, the task involves elements of both recognition and recall. So, although the use of a PCA database and an array-based interface can be argued to overcome the problems associated with verbal descriptions and working featurally, they will still necessitate the use of recall, albeit it in a restricted form.

The design of PCA facial compositing systems therefore *appears* to be a step in the right direction in terms of a good fit between technology and cognition, although early tests of images made with EVOfit found them to have poor utility (Frowd, Hancock, and Carson 2004) and that E-FIT and sketch artist images were generally superior (Frowd, Carson, Ness, Richardson, Morrison, McLanaghan, and Hancock 2005). Nonetheless, a key issue may be that array-based interfaces do away with the need for lengthy and complex interactions between the witness and the system via the operator. One point to note here is that, although it is often very difficult and can undoubtedly lead to poor results, witnesses required to use feature-based compositing systems are generally able both to communicate with the operator and to describe what they want the system to do. A key question, therefore, is whether the problems associated with earlier systems (such as E-FIT) were due to the fact that they involved *any* recall and verbalisation or because the witness was *forced* into using recall and verbalisation throughout the construction process. Could it be that at least some of the information that the witness wants to provide is both accurate and useful? If so it could be argued that by not being able to incorporate this information, array-based interfaces are missing out on potentially vital cues and are in fact not interacting as well as they could with the cognitive processes of the witness.

The present paper describes a study that was designed to look at the above questions in more detail. In particular, the interaction between the technology involved in array-based facial compositing systems and witness cognition was examined. This was achieved by asking witnesses to interact with an array-based system and recording their reactions and comments. Rather than make use of an actual PCA compositing system, arrays of faces were created in advance. These arrays were arranged sequentially so that the images progressed towards a good likeness of the target face (although not a perfect 'picture' of the target, as this would be forensically unrealistic). This was done so that each participant worked through exactly the same images, making their comments directly comparable, and so that the images displayed did actually end up resembling the target face. This technique meant that the participants were not actually affecting the construction of the composite, although they did think that they were, and therefore allowed the interaction to be observed under more controlled circumstances. Critically, it allowed the amount and type of variation between the faces in the array to be controlled and pre-determined. In addition, the individual faces used in the arrays were constructed using the E-FIT compositing system, rather than a PCA system. Although this does mean that the arrays and images used differ from those that would be generated by a PCA system, the method adopted allowed the separate manipulation of features and the configuration of features and also provided semantically meaningful differences between the arrays, for example, the faces in one array may have shared five features or had just one feature in common.

As well as the amount of variation between the faces in each array, two other factors were examined: the number of arrays that the participants were asked to work through was manipulated to create a shorter sequence of 30 arrays and a longer sequence of 60 arrays; and the task the participant was asked to perform was manipulated so that some participants were asked to select the best match for the target from each array, whilst others were asked to select the most masculine face from each array. This latter task was introduced because by requiring participants to simply select the most masculine face from each array, any active recall of the target face and comparison with the faces in the arrays was removed.

# 2. Method

## 2.1 Design

This experiment employed a mixed design examining the within-participant factors of sequence length, with two levels (short, 30-array sequence; long, 60-array sequence) and array type, with six levels (five features changed; four features changed; three features changed; two features changed; one feature changed; configural changes only) and the between-participant factor of task type, with two levels (best match for the target selection or most masculine selection). Two target faces were used and the design was fully-crossed and counterbalanced for both array sequence order and target face order. As the study was exploratory, many dependent variables were included (and are detailed in the results and discussion), such as both quantitative and qualitative feedback on the array interface and process in general, consistency in selections both between and within participants and performance on a line-up task designed to test memory for the target face.

## 2.2 Materials: Target photographs

Two target 'suspect' photographs were used as stimuli, each showing a full-face head-and-shoulders view of a person unknown to the participants. The targets were Caucasian middle-aged males, who were shown clean-shaven, with no spectacles or other facial paraphernalia and no distinguishing marks. Colour images of both targets were presented in high resolution on a 17" computer screen.

## 2.3 Materials: Images, arrays and sequences

For each target an 'optimum' likeness was produced using E-FIT v3.1a by an operator with several years of experience with the E-FIT system and checked for accuracy by a panel of five trained operators. These composites formed the 'base image' for each target, from which the subsequent E-FIT images were derived as detailed below.

In order to produce an image sequence that would develop a progressively closer likeness to the target, the 'base image' was gradually modified using the E-FIT system to become increasingly *less* like the target. These modifications ranged from small changes in facial configuration (the smallest differences, which should remain fairly good type-likenesses of the target) to entirely new facial images sharing no features with the base-image. Placed in reverse order these images would therefore produce a sequence running from entirely dissimilar, effectively randomly-generated images bearing little or no likeness to the target (other than by coincidence) through progressively more similar-to-target images until finally reaching the optimum base image itself.

Five sets of each array type were created, with nine images being created for each set. This gave six 'types' of image, according to the number of features on which they differed from the optimum. Each type had five sets of images, differing in terms of which features were changed for the one, two, three, four and five-features changed types, and the amount of feature displacement for the configural change type. Each set consisted of nine individual images, giving a total of 270 images for each target. The nine images within each set were cropped and resized to an approximately uniform size and positioned in a Microsoft PowerPoint slide to form a 3 × 3 array.

The arrays of images for each target were then arranged in PowerPoint slide sequences such that the *least*-like image arrays (sharing none or few of the optimum feature exemplars) would appear towards the beginning of the sequence and the *most*-like image arrays (sharing most or all of the optimum feature exemplars) would appear towards the end of the sequence. When run, the sequence therefore proceeded from composites bearing little or no resemblance to the target, through composites bearing progressively more of a resemblance to the target, and ending with composites bearing a good resemblance to the target. However, as it was envisaged that a composite system operating on a genetic algorithm would be expected to occasionally take a backwards step, there was some mixing of the image types in the array sequence to simulate such backwards steps. Specifically, the all features change and four-changes arrays were intermixed, the three-changes and two-changes arrays were intermixed, and the one-change and the configural-changes arrays were intermixed.

For the long sequence condition for each target (60 arrays) the same overall pattern of array presentation was maintained as was used in the short sequence; however, each set of images appeared twice in the sequence. For the second appearance of an image set, the images were pseudo-randomly rearranged so that they appeared in a different position in their second array than they had in their first. Thus, for example, an image which had appeared in the centre of the array on its first appearance in the sequence might appear in the top-left of the array on its second appearance. The occurrence of the repeated image sets was pseudo-randomised, with the restriction that an array pair was never presented consecutively. For both the short and long sequences of arrays, the final array shown consisted of images with the smallest configural changes (single-feature, single-pixel displacements) including the optimum base-image itself.

## 2.4 Materials: Photospread line-ups

For each target a target-absent and a target-present photospread line-up was created. The target-absent photospreads consisted of nine foil images, whilst the target-present photospreads consisted of an image of the target altered to be different to that seen previously (lighting, background, clothing and hair cues were changed) and eight foils (different from those used for the target-absent photospread). The foils for both photospreads were judged to resemble the target visually as well as to verbal descriptions provided of the targets by participants who were unfamiliar with them. The foil images were partly sourced from the Pics image database maintained by Stirling University (http://pics.stir.ac.uk). All of the images for the photospreads were standardised so as to be pictorially similar.

## 2.5 Participants

60 Open University staff volunteered as participants in this study. There were 47 females and 13 males, with ages ranging from 23–55 years. None were familiar with either of the targets or any of the photospread line-up foils.

## 2.6 Procedure

Participants were briefed as to the nature of the research and those in the 'best match' task condition informed that their role would be to act as a participant witness and study the face of a target 'suspect' for 30 seconds before working through a sequence of arrays of faces on the computer. They were also instructed that in each array they should select the face that was most like the target and, based on their selections, the computer should progress towards an improved likeness of the suspect. Those in the 'most masculine' condition were told simply that they would need to select the most masculine from each array. The participant was told to study each array of nine face-images, and decide either which bore the most resemblance to the suspect or which was the most masculine (depending on condition). Participants were also encouraged to verbalise their thoughts as much as possible when working through the arrays, which, along with their decisions, were recorded by an experimenter. In the 'best match' condition, when the final array was reached the participant was also asked to rate on a 10-point scale how good a likeness of the suspect the chosen image was (whilst it was still in view). The participant was then shown first a target absent and then a target present nine image photospread line-up and told that the target suspect may or may not be present before each one. As well as indicating whether the target was present, and if so which image was of the target, the participant also gave a rating of their confidence in that decision using a 10-point scale. After completing the line-up task, the participant then worked through the sequence of arrays for the other target, following exactly the same procedure. The final stage of the study involved asking the participant questions about their experience of the arrays and eliciting any further comments or observations that the participant wished to make. Following this free-commentary session the participant was thanked for their assistance and de-briefed.

## 3. Results and Discussion

### 3.1 Participant feedback from best match selection conditions

The first thing to note is that all of the participants managed to work through both the 30 and 60 array sequences and to select a 'best match' in each array, with no participants giving-up due to task difficulty (or any other reason). This demonstrates that selecting the best match from an array of faces is at least something that the participants seemed able to do.

The verbalisations made by each participant were analysed to discover any common themes and to determine the frequency with which certain comments were made. This analysis revealed that in 56.25% of the sequences (each participant worked through two sequences, one short and one long), the participant commented in response to at least one array that they found the task of selecting the best match to be difficult and in 75% of sequences the participant commented for at least one array that they thought that none of the faces in the array looked sufficiently like the target to be selected as a 'best match'. Analysis also revealed that participants quite often wanted to interact with the array in a different manner from simply choosing the best match. For example, in 28% of sequences the participants responded that they would prefer to indicate the face(s) that looked the least like the target, rather than the most; in 27.5% of sequences comments were made about wanting to select multiple faces from an array rather than just one; and in 70% of sequences the participants said they would like to alter or select individual features, rather than the whole face. Further analysis revealed that participants often felt that they had become confused, with mention being made in 31.25% of sequences that the participant was having difficulty remembering what the target face looked like.

One of the standardised questions asked after the array task concerned the length of the sequences. Analysis revealed that 60% of participants thought that 30 arrays was 'the right number' to work through and further that 58% of participants believed that 60 arrays was 'far too many' to work through. This is an important figure, as one potential advantage of PCA based systems is that they can produce a credible likeness much more quickly than feature-based systems. The data collected here suggest, though, that the repetitive nature of the task means that witnesses may grow fatigued or disillusioned unless a good likeness can be reached within about 30 generations, even though this probably involves less than a third of the time it takes to construct a composite using systems such as E-FIT and PROfit. Informal qualitative analyses of the comments made by participants suggest that being presented with screen after screen of similar looking faces can make it difficult to remember what the target face looked like. It is also likely that requiring witnesses to make relative judgements about which of the faces presented is the best match involves more than the cognition required to decide if a presented face is, or is not, the target. The use of relative judgements of simultaneously presented faces has been shown to be problematic in the psychological literature on eyewitness identification, as it tends to lead witnesses to make misidentifications by choosing the best match for the perpetrator, even if the best match is not actually a particularly good match. Although selecting the best match is the aim of the array-based interface, the participants' responses suggest that this can become problematic if a large number of arrays are required to produce the final composite.

The participants were also asked about the selection method employed, i.e., selecting the best match and other potential selection methods. 58% of participants responded that choosing the single best image was 'frequently' difficult, with 5% saying it was 'always' difficult and none saying it was 'never' difficult. Opinion proved to be more divided when they were asked whether they would have preferred to use a selection method based on choosing the best two or three images in each array, with 35% responding 'probably' and 35% responding 'probably not'. A third type of selection method, that of providing a score out of ten for each of the faces in the array, has the potential to provide the system with a great deal of information, as a response will be given to each of the nine faces. This would allow each to be weighted differentially when generating the next array and might allow the genetic algorithms employed in PCA systems to converge on the target more swiftly. However, 42% of participants responded that they would 'probably not' have preferred to use this method and 40% said that they would 'definitely not' have preferred to use it.

The final question about selection methods concerned giving specific feedback about the array, so that the participant could respond however they wished and the system would be able to accommodate their comments. With 75% of participants responding that they would have 'definitely' preferred this type of interaction with the system and a further 20% saying they would 'probably' have preferred it, this selection method proved to be the most popular. Systems such as E-FIT were designed so that a trained operator would be able to take virtually any comment made by the witness and translate it into a useful alteration to the composite being constructed. However, it is exactly this form of verbalisation that some researchers have suggested is a potential source of error in composite construction and is one that array-based interfaces can be designed to avoid if they employ the best-face selection method. The point is, though, that whilst the interface used by E-FIT *requires* the witness to verbalise *every* aspect of their decisions, if greater flexibility of selection were built into an array-based interface the witness would not *have* to verbalise every aspect of selection, but could do so if they desired.

## 3.2 Individual differences in faces selected from arrays

As well as examining the interaction between technology and cognition, the experiment conducted also allows an examination of individual differences in the cognitive processes employed. One benefit of the methodology employed in the experiment is that each participant was presented with exactly the same arrays. It was therefore possible to see whether participants tended to select the same face as the best match for the target (or the most masculine face) or whether different participants selected different faces even though they were presented with exactly the same images in exactly the same order. Logically, it would be expected that there would be greater consistency in the faces selected from arrays containing a large degree of variation, such as those where the faces differed on four or five features, than those containing less variation, such as the arrays containing faces sharing all but one feature or those where only the configuration of the features was different.

The frequency with which each face in each array was selected as either the best match to the target or the most masculine face was calculated and these responses were then grouped according to the number of features changed in the array (i.e., five, four, three, two or one feature or just configural changes). From these data it was possible to calculate the mean percentage of participants who selected the modal (most selected) face. A summary of these data is presented in Table 1.

These data were subjected to a 2 (task type — best match or masculinity selection) × 2 (length — 30 or 60 arrays) × 6 (number of features changed — 5 features through to configural only changes) between-participants[1] ANOVA which revealed: a statistically significant main effect of task type ($F(1,156) = 15.067; p < 0.001$), with an effect size (partial eta$^2$) of 0.091; a statistically non-significant main effect of sequence length ($F(1,156) = 0.825; p = 0.365$), with an effect size of 0.005; a statistically non-significant main effect of array type ($F(5,156) = 1.886; p = 0.1$), with an effect size of 0.057; statistically non-significant two-way interactions between task type and sequence length ($F(1,156) = 0.2797; p = 0.598$), effect size of 0.002, task type and array type ($F(5,156) = 0.645$; $p = 0.695$), effect size of 0.02, and between sequence length and array type ($F(5,156) = 0.774$; $p = 0.57$), effect size of 0.024; and a statistically non-significant three-way interaction ($F(5,156) = 0.185$; $p = 0.968$), effect size of 0.006.

The only statistically significant result was the main effect of task type, with participants generally showing more agreement about which face was the most masculine (mean of 28.22% of participants choosing the most selected face) than about which was the best match to the target (mean of 24.5% of participants choosing the most selected face). By chance alone, each face would be picked by 11.11% of participants, so the means from both the masculine and best match selection conditions are more than twice as much as would be expected by chance. However, on average approximately only a quarter of participants selected the most chosen face, suggesting there are considerable individual differences in the perception of faces; whether this be in determining masculinity or the task of comparing the faces to the memory of the target.

From Table 1 and the results of the inferential analysis it is apparent that there was a great deal of variation in the faces selected as the best match. In addition, there was not a great deal of difference in this variation according to the similarities of the faces within the arrays, with 28% of participants selecting the modal face on average in the five-features changed arrays and 24.5% on average in the one-feature changed array. In many ways this result is surprising, as the task of selecting the best match from an array of very different looking faces should be far easier (and should therefore lead to fewer individual differences) than selecting the best match from an array of very similar looking faces. One possible explanation is that the participants had simply forgotten what the target looked like, but a check of their memory following the sequence of arrays revealed that their memories appeared intact (see later analyses). Instead the results suggest that there are

**Table 1.** Percentage of participants selecting 'most chosen' face, by condition, sequence length and number of features on which faces differed

| | | | No. of features changed | | | | | |
|---|---|---|---|---|---|---|---|---|
| | | | Five | Four | Three | Two | One | Configural |
| Best match | Short sequence | Mean | 28 | 26 | 26.5 | 21.5 | 24.5 | 22 |
| | | SD | 8.91 | 6.75 | 3.79 | 3.79 | 3.71 | 3.26 |
| | Long sequence | Mean | 26.25 | 26.5 | 24.75 | 23.75 | 24.75 | 20.25 |
| | | SD | 6.37 | 5.43 | 5.2 | 6.26 | 7.02 | 3.22 |
| Most masculine | Short sequence | Mean | 34 | 30 | 27 | 26 | 29 | 29 |
| | | SD | 8.9 | 5 | 5.7 | 4.18 | 7.41 | 12.45 |
| | Long sequence | Mean | 28 | 28 | 26.5 | 29 | 28 | 27 |
| | | SD | 7.45 | 6.75 | 5.79 | 4.59 | 7.53 | 5.37 |

considerable differences in the way that faces, at least unfamiliar faces, are encoded, stored and retrieved from memory.

That there are such large individual differences in face processing cognition is important in considering the design of facial compositing technology. Perhaps most importantly there is a need to consider whether the aspects of the face that the participant is concentrating on to make their decision about which is the best match, are similar to the aspects of the face that the system will be using to generate the next array. For example, the large individual differences could have resulted from some participants selecting the best match because they though the face-shape was a good match, some because the eyes were a good match and some because the overall look of the face was a good match for the target. If the compositing system is based on a PCA analysis of the entire face, then the next array generated may well exclude the aspects of the face that the participant concentrated on, especially if these were particular individual features. It could well be that a more flexible system, one that includes technology that can respond more directly to the cognition of the witness, by allowing a specific feature to be used prominently in generating the next array for instance, would be better able to cope with the obvious large individual differences that exist.

## 3.3 Within-participant consistency in faces selected from arrays

As well as allowing an examination of individual differences in the interaction of cognition and technology, the design of the present study also allowed consistency *within* the selections made by each participant to be investigated. This was only possible in the longer (60 array) sequences where each array appeared twice, so that the same faces were seen but in different positions within the array. It was therefore possible to determine how often each participant selected the same face from these matched arrays and how often they chose a completely different face, even though the face they selected the first time the array was presented was still available.

These data are presented in Table 2, and as before grouped according to whether they related to arrays that had either five features, four features, three features, two features, one feature or just configural changes.

The data were subjected to a 2 (task type — best match or masculinity) by 6 (array type — from 5 to configural changes) mixed design ANOVA, which revealed: a statistically non-significant main effect of task type ($F(1,58)=0.802$; $p=0.374$), with an effect size (partial eta$^2$) of 0.014; a statistically significant main effect of array type ($F(5,290)=13.528$; $p<0.001$), with an effect size of 0.19; and a statistically non-significant interaction between the two factors ($F(5,290)=0.375$; $p=0.865$), with an effect size of 0.006.

**Table 2.** Mean percentage of matched array pairs from which the same face was selected, by array type

| | | No. of features changed | | | | | |
|---|---|---|---|---|---|---|---|
| | | Five | Four | Three | Two | One | Configural |
| Best match | Mean | 47 | 39 | 34 | 30.63 | 22 | 21.37 |
| | SD | 27.76 | 25.2 | 21.82 | 27.2 | 18.56 | 18.95 |
| Most masculine | Mean | 53 | 47 | 32 | 31.5 | 26 | 23 |
| | SD | 21.79 | 22.73 | 20.93 | 27 | 19.58 | 26.92 |

From an examination of Table 2 and the results of the inferential analysis, it is apparent that participants were generally more consistent when there was greater variation present in the array (five or four feature changes) than when there was less (one feature or configural changes). In other words, when there was greater variation a participant was more likely to select the same face from the two presentations of the array than when there was less. However, the data still suggest that there is considerable variation in the selections made by the same participant, as even in the five features changed condition over half the participants selected a different face the second time they were presented with the array and this figure rose to approximately 75% of participants making inconsistent selections in the configural changes condition.

The above analysis also shows that the best match and masculinity selection conditions produced similar results, as the main effect of task type and its interaction with array type were statistically non-significant. These results could suggest that the order with which the faces are presented in the array is important, as this order was changed between presentations of the matched arrays and most participants tended to select a different face. If participants were working sequentially through the faces, starting say with the top-left and finishing with the bottom-right, it could be that the serial order in which the face appeared was affecting its selection.

An alternative explanation is that the arrays seen between presentations of the two matched arrays were affecting the selection made. This could either be because the memory for the target face was being changed by exposure to the faces in the array or because the decisions made to the intervening arrays were leading the participant to concentrate on an aspect of the faces that led to a different face being selected from the second presentation.

Whatever the actual mechanism, this result suggests that the design of compositing technology could well benefit from allowing a witness to make a selection based on several faces rather than one; a suggestion that was made quite frequently by participants in this experiment (see analysis of feedback above). This would allow them to incorporate aspects from several faces if they felt that more than one face had something in common with the target. In addition, the results suggest that narrowing the amount of variation present within the arrays too greatly could be problematic, particularly if participants focus on one aspect of the face at a time. For example, if they first focus on getting the face-shape correct and then move onto the internal features, the arrays will need to contain sufficient variation so that a suitable selection of features is displayed. Without this variation the participant will be stuck with elements of the face that they had not actively selected.

## 3.4 Memory for the target face

Analysis of the feedback received from participants revealed that in nearly a third of all the sequences (31.25%), the participants reported having difficulty remembering what the target face looked like. Obviously, if their memory for the target face has been eroded in some way, either by the passage of time or by viewing the faces presented in each array, the task of composite construction will become very difficult indeed, if not impossible. However, it is possible that the participants' comments were more a reflection of the difficulty they were having in actively recalling the target face than any deterioration in the actual stored memory of the face *per se*. Although problems in recalling the face are far from trivial, they are not nearly as limiting as would be the case if the process of working with the array interface actually affected the stored memory.

To determine whether working through the arrays had indeed altered or eroded the memory for the target face, each participant's memory for the face was tested after they had completed each sequence. First of all they were asked to study a photo line-up consisting of nine faces, which (unknown to them) did not contain an image of the target, to say whether the target was present and if so, to indicate the appropriate image. They were then shown a second photo line-up containing different faces, one of which was the target, and asked for the same decisions. A summary of the data relating to how accurately these tasks were performed is presented in Table 3.

As can be seen from Table 3, the vast majority of participants were both able to determine correctly that the target was not present in the target-absent line-ups and to pick-out the target from the target-present line-ups. To complete this task, the participant's memory of the target's face must have been reasonably intact, and not particularly altered by exposure to the faces in the arrays.

Although performance across task type was similar in the case of the target present conditions, more mistakes were made with the target absent line-ups in the masculine selection condition than in the best match selection condition. Analysis of the data using the chi-square procedure (making use of the Fisher's exact test

**TABLE 3.** Outcome (in %) of photospread line-ups by condition (TA = Target Absent; TP = Target Present)

| | | | Hit | Miss | False Alarm | Correct rejection |
|---|---|---|---|---|---|---|
| Best match | Short sequence | TA | na | na | 0 | 100 |
| | | TP | 90 | 10 | 0 | na |
| | Long sequence | TA | na | na | 2.5 | 97.5 |
| | | TP | 87.5 | 12.5 | 0 | na |
| Most masculine | Short sequence | TA | na | na | 5 | 95 |
| | | TP | 90 | 10 | 0 | na |
| | Long sequence | TA | na | na | 15 | 85 |
| | | TP | 90 | 0 | 10 | na |

to compensate for expected counts less than 5 and using Cramer's V as an estimate of effect size) revealed that the difference between best match and masculine selection conditions was statistically significant for the target absent line-ups ($chi^2 = 5.11$; df = 1; $p < 0.05$; effect size = 0.206), but statistically non-significant for the target present line-ups ($chi^2 = 0.43$; df = 1; $p = 0.835$; effect size = 0.019).

Thus there appears to be some advantage to interacting with the arrays in a manner that requires constantly accessing the memory for the original target. The fact that the difference was significant for the target absent and not the target present parades is probably due to the target absent parades presenting a more difficult task; as it is often easier to decide that the target face is present than to decide that none of the faces is that of the target. Moreover, it is possible that seeing the faces in all the arrays but *not* continually accessing the memory for the target face had a more detrimental effect on memory for the target than resulted from the constant recall of the target in the best match conditions. The main point here is that whatever negative effect exposure to the arrays seems to have on memory for the target appears to be overcome if the target is recalled and compared to the faces in the arrays.

## 4. Conclusions

The main conclusion to be drawn from the current study is that it could be inadvisable to limit the interaction of a witness/participant with array based facial compositing technology to making just a single selection from each array. Instead, the large differences between individual participants (and also within a single participant throughout the process) suggest that greater flexibility is required in order to arrive at a good likeness. In particular, participants want to be able to respond to multiple faces in each array and also to specific aspects (such as individual features) of certain faces.

These findings appear at odds with the wider literature that suggests that both witness recall and verbalisation are important sources of error when constructing a facial composite. However, the studies these suggestions were based on made use of compositing technology that forced the witness to *always* verbalise their thoughts and to rely very heavily on recall. It could be that involving a greater amount of cognition based on recognition and allowing the witness to provide more specific comments only when they want to and only about those aspects that they want to cover, could lead to a more efficient and accurate compositing system.

## Notes

* This research was conducted with the support of EPSRC grant GR/S06745/01(P). The authors would like to acknowledge the kind assistance of Dr Chris Solomon and the Forensic Imaging Group at The University of Kent.

1. The analysis is entirely between-participants as it is conducted by-item, i.e., with the arrays rather than the participants acting as the item of analysis.

## References

Brace, N.A., Pike, G.E., Allen, P., and Kemp, R.I. In press. "Identifying composites of famous faces: Investigating memory, language and system issues". *Psychology, Crime and Law.*

Buckhout, R. 1974. "Eyewitness testimony". *Scientific-American* 231(6): 23–31.

Christie, D.F. and Ellis, H.D. 1981. "Photofit constructions versus verbal descriptions of faces". *Journal of Applied Psychology* 66(3): 358–363.

Cutler, B.L. and Penrod, S.D. 1995. *Mistaken Identification: The Eyewitness, Psychology, and the Law.* New York: Cambridge University Press.

Davies, G., van der Willik, P., and Morrison, L.J. 2000. "Facial composite production: A comparison of mechanical and computer-driven systems". *Journal of Applied Psychology* 85(1): 119–124.

Ellis, H.D., Davies, G.M., and Shepherd, J.W. 1978. "A critical examination of the Photofit system for recalling faces". *Ergonomics* 21(4): 297–307.

Fahsing, I.A., Ask, K., and Granhag, P.A. 2004. "The man behind the mask: Accuracy and predictors of eyewitness offender descriptions". *Journal of Applied Psychology* 89(4): 722–729.

Frowd, C.D., Carson, D., Ness, H., Richardson, J., Morrison, L., McLanaghan, S., and Hancock, P. 2005. "A forensically valid comparison of facial composite systems". *Psychology, Crime and Law* 11(1): 33–52.

Frowd, C.D., Hancock, P.J., and Carson, D. 2004. "EvoFIT: A holistic, evolutionary facial imaging technique for creating composites". *ACM Transactions of Applied Perceptions* 1(1): 19–39.

Gibling, F. and Bennett, P. 1994. "Artistic enhancement in the production of photo-Fit likenesses: An examination of its effectiveness in leading to suspect identification". *Psychology, Crime and Law* 1: 93–100.

Gibson, S.J., Solomon, C.J., and Pallares-Bejarano, A. 2003. "Synthesis of photographic quality facial composites using evolutionary algorithms". In R. Harvey and J.A. Bangham (eds), *Proceedings of the British Machine Vision Conference 2003* 1: 221–230.

Hancock, P.J. 2000. "Evolving faces from principal components". *Behavior Research Methods, Instruments and Computers* 32(2): 327–333.

Hancock, P.J.B., Burton, A.M., and Bruce, V. 1996. "Face processing: Human perception and principal components analysis". *Memory-and-Cognition* 24(1): 26–40.

Laughery, K.R. and Fowler, R.H. 1980. "Sketch artist and Identi-kit procedures for recalling faces". *Journal of Applied Psychology* 65(3): 307–316.

Leippe, M.R. and Wells, G.L. 1995. "The police lineup: Basic weaknesses, radical solutions". *Criminal Justice and Behavior* 22(4): 373–385.

Levi, A.M. and Jungman, N. 1995. "The police lineup: Basic weaknesses, radical solutions: Reply". *Criminal Justice and Behavior* 22(4): 386–396.

O'Toole, A.J. and Thompson, J.L. 1993. "An X Windows tool for synthesizing face images from eigenvectors". *Behavior Research Methods, Instruments and Computers* 25(1): 41–47.

Pozzulo, J.D and Warren, K.L. 2003. "Descriptions and identifications of strangers by youth and adult eyewitnesses". *Journal of Applied Psychology* 88(2): 315–323.

Rakover, S.S. 2002. "Featural vs. configurational information in faces: A conceptual and empirical analysis". *British Journal of Psychology* 93(1): 1–30.

Shepherd, J.W., and Ellis, H.D. 1996. "Face recall — methods and problems". In S.L. Sporer (ed), *Psychological Issues in Eyewitness Identification*. Mahwah, NJ: Lawrence Erlbaum Associates, 87–115.

Tanaka, J.W. and Farah, M.J. 1993. "Parts and wholes in face recognition". *Quarterly Journal of Experimental Psychology: Human Experimental Psychology* 46A(2): 225–245.

Wells, G.J. and Seelau, E.P. 1995. "Eyewitness identification: Psychological research and legal policy on lineups". *Psychology, Public Policy and Law* 1(4): 765–791.

# Perceptual recalibration in sensory substitution and perceptual modification

Juan C. González, Paul Bach-y-Rita and Steven J. Haase
Universidad Autónoma del Estado de Morelos / University of Wisconsin / Shippensburg University

This paper analyzes the process of perceptual recalibration (PR) in light of two cases of technologically-mediated cognition: sensory substitution and perceptual modification. We hold that PR is a very useful concept — perhaps necessary — for explaining the adaptive capacity that natural perceptive systems display as they respond to functional demands from the environment. We also survey critically related issues, such as the role of learning, training, and nervous system plasticity in the recalibrating process. Attention is given to the interaction between technology and cognition, and the case of epistemic prostheses is presented as an illustration. Finally, we address the following theoretical issues: (1) the dynamic character of spatial perception; (2) the role of functional demands in perception; (3) the nature and interaction of sensory modalities. We aim to show that these issues may be addressed empirically and conceptually — hence, the usefulness of sensory-substitution and perceptual-modification studies in the analysis of perception, technologically-mediated cognition, and cognition in general.

## 1. Sensory substitution

For the brain to correctly utilize information from sensory substitution devices, it is not necessary that it be presented in the same form as in natural sensory information systems. We do not *see* with the eyes; the visual image does not go beyond the retina, where it is transformed into patterns of pulses along nerves. It is the *brain* which generates representations from the patterns of pulses as the person interacts with the environment. This has led to the development of tactile vision sensory substitution (TVSS) systems to enable information from an artificial sensor (e.g., TV camera) to be delivered to the brain (Bach-y-Rita et al. 1969). After training with the TVSS, blind subjects using the TVSS system report attention being drawn away from the skin on which the interface is placed and redirected outwardly to the distal objects in three-dimensional space (if they control the camera movement). Subjects learned to make perceptual judgments using visual means of analysis, such as perspective, parallax, looming and zooming, and depth judgments. They have also learned to identify complex objects such as faces and body position, and have performed real-time hand-"eye" coordination tasks such as batting a ball and performing miniature electronic component manufacture (reviewed in Bach-y-Rita 1972, 1995; Bach-y-Rita et al., in press).

As is the case with other epistemic prostheses, the TVSS system endows the agent with a supplementary cognitive capacity that is critically dependent on a previous period of active learning. During this period, the agent's perceptual learning consists of activity correlating sensory inputs with motor responses so as to ensure adequate behavior (i.e., a behavior that simultaneously establishes sensorimotor invariants and satisfies functional needs). As will be seen, it is precisely this late-learning perceptual process that can be interpreted in terms of a *recalibration* of the perceptual system. For the time being, let us observe that in order for the learning process to be effective, productive and lasting, it has to satisfy real functional demands; that is to say, the perceptual abilities acquired must have a useful application to the behavior or cognitive performance of the agent.

A recently developed brain-machine interface (BMI) through the tongue offers the potential for a practical sensory substitution device. The tongue system comprises a small video camera and a computerized control box, which translates the spatial information into mild electric impulses, interfaced with a ribbon of wires leading to a square of 144 gold-plated electrodes held against the subject's tongue (Bach-y-Rita et al. 1998, in press).

From a philosophical perspective, the fact that human beings successfully perform — and even have some qualitatively-similar perceptual experiences — in the contexts of both sensory substitution and (as we will see later) perceptual modification carries some interesting implications. One of these implications is captured by a concept that we call 'perceptual recalibration' (PR). In what follows we will elaborate on such a concept in order to better understand the nature of perception and aided cognition, and to bring forth certain issues that critically relate to PR.

## 2. Perceptual recalibration

PR refers to the process of successful adaptation of a natural and multimodal perceptual system when its normal operation is undermined or modified (by accidental damage or voluntary modification of one of its modalities, for instance) but recovers its capacity to operate effectively as judged by its new overall performance. This adaptation relates to both the sensory and the cognitive levels involved in perception (i.e., to the intake and utilization of outer information), but also to proprioception and perhaps other forms of internal awareness. We will focus here mainly on perceptual processes. The concept of PR is illustrated in the successful behavior of a blind person walking with the aid of a cane or in the successful operation of a remote-controlled lunar vehicle. The point in talking about 'recalibration' (a concept that has been already used by psychologists like James J. Gibson (1966) and philosophers like Fred Dretske (1995) and Joëlle Proust (2000)) is to suggest that the adaptive process that takes place following impairment or modification of a perceptual system should be understood in terms of the background of the *calibration* that naturally relates functional demands, perceptual performance, and appropriate motor responses. From this angle, the capacity to recalibrate depends on the possibility to calibrate. Let us first turn to the issue of *recalibration*.

We use the term 'recalibration' in Dretske's sense (1995) when he refers to the adjustment or adaptation that a sensory system is capable of. Dretske (1995: 20ff) mentions the case of visual modification prisms that displace the retinal image to the left by 30°. After a certain time and interaction between the visual and the tactile systems, it is possible to recalibrate the tacto-visual system so as to successfully adapt to these new circumstances.

In the case of the TVSS, the stage of successful adaptation is demonstrated by the fact that one can (after learning) change the location of the artificial eye (i.e., the camera) and/or the location of the dermal interface without compromising the agent's performance. Indeed, as long as the agent maintains motor control of the sensory-input device, it is possible to change its location (say, from the head to the hand or even to the end of a hand-held pole) and/or the location of the interface (say, from the back to the abdomen), and still guarantee an adequate level of perceptual performance (Bach-y-Rita 1972, 1995). This is a dramatic demonstration of the importance of the agent's endogenous movement and the establishment of sensorimotor invariants in the constitution of perceptual spaces.

## 3. Perceptual modification

Together with sensory-substitution studies, perceptual-modification experiments provide a very attractive case to understand PR. Perhaps the clearest and most concise way to explain why these experiments are so attractive is by quoting Ivo Köhler, when he tells us that "modification illuminates development" (Dolezal 1982: 14). Although there is a rather large spectrum of perceptual-modification experiments, we will concentrate on a classical example: the use of visual-modification lenses by human subjects.

Inspired by Stratton's (1897) experiments with modification lenses, many other tests aimed at (dis)proving the adaptive capacity of the perceptual system were carried out throughout the 20th century, giving rise to a number of hypotheses and explanatory theories that those experiments intended to confirm.[1] Among these experiments, we can cite those of Gibson (1966), Köhler (1964), Harris (1965), Held (1965), Rock (1966), Rock & Harris (1967) and Dolezal (1982).

Visual-modification lenses typically disturb the path of light reaching the eye so as to displace, invert, fragment, tilt, reverse, reduce, augment or distort the retinal image. The overall aim of those modifications is to alter the sensorimotor invariants established during the agent's development, thereby disturbing the cognitive relation that naturally holds between the agent and its environment. The study of those controlled modifications can help evaluate the degree of perceptual adaptation to a new situation. The criteria for evaluating the degree of adaptation typically include at least three levels: (a) the phenomenal/qualitative experience of the agent (the way things look or feel); (b) the agent's verbal reports; (c) the agent's behavior and interaction with the environment. Therefore, before we can speak of a successful or failed adaptation, we should distinguish the level or levels to which we are applying our criteria of evaluation.

One undeniable and recurrent fact that appears in the conclusions of the above experiments is that a perceptual recalibration (manifested in some degree of adaptation as judged by at least one of the mentioned criteria) virtually always occurs. This is also confirmed by previous experiments (Ewert 1930; Snyder & Pronko 1952; Köhler 1951; Kottenhoff 1957). Hence it is obvious that we as human beings have the capacity to adapt, within certain limits, to abnormal or new perceptual circumstances – something that a great deal of animal species seem to lack (Vurpillot 1963; Gregory 1981: 209ff). However, the exercise of the said capacity depends on a stage of active learning and on subsequent practice — typically through the fulfillment of an actual functional demand.

## 4. Calibration and learning

As for the issue of *calibration*, it is convenient to distinguish two related though different aspects of it: the *capacity* of natural perceptual systems to be calibrated (which seems to be explainable phylogenetically), and *the actual calibration* of said systems (which seems to be explainable ontogenetically). Let us start with the latter.

If the successful adaptation of the perceptual system is conceived in terms of a process of *recalibration*, it seems legitimate to posit that the *calibration* of the system consists in the establishing of stable and lasting sensorimotor invariants during the early stages of the agent's perceptual experience (typically during infancy and, at least in some species, within critical temporal windows of neuronal maturation).[2]

In the normal course of human cognitive development, the perceptual systems are naturally calibrated through the feedback process of perceptual learning that a person will have undergone since infancy. Berkeley (1709) proposed something very similar, as did Piaget (1952) in the mid 20th century. More recently, numerous studies with infants have demonstrated that the learning process allows sensory input to be associated with relevant motor responses so as to ensure the satisfaction of increasingly complex functional demands. Thus, for example, self-feeding in infants demands previous mastering of the grasping of objects, which in turn involves the coordination between target-perception and accurate movement of bodily parts (McCarty et al. 2001; McKenzie et al. 1993; Morrongiello & Rocca 1989) — the whole process being underlain by the need to achieve an explicit goal (e.g., *grasping a cookie*).[3] Perceptual calibration can thus be conceived as the process that allows the feedback loop involving perception, goal and action to be established and become stable around functional demands.[4] Hence, in regard to perception, learning amounts to calibrating — an idea that garners support in other areas of human perception and cognition such as reading and word recognition via implementation of parallel distributed processing (PDP) networks (McClelland & Rumelhart 1986).

Now, in respect to the system's *capacity* to be calibrated, it is clear that the innate physiology of the system itself plays a major role. Many human perceptual limitations have been explained in terms of the nervous system's hard-wiring, and differences in perceptual learning across diverse animal species may be related to their innate nervous system architecture. For example, the fact that hens have small brains may account for their behavior while wearing distortion prisms (that shift the distal object's image, say, 7° to the right), for they will not adapt (recalibrate) and will keep pecking to the side of the grains (Gregory 1981: 210)!

Thus, it is reasonable to assume that the innate physiology and anatomy of the organism determines, to a great extent, the limits of what can be perceptually processed and acted upon and hence, of what can be learned.[5]

The absence of change in pecking behavior by hens and other cases of lesser or nonexistent learning capacity in non-human living beings is undoubtedly related to brain mass (Bach-y-Rita & Aiello 2001), but other factors are certainly important. One of these, as we will see, is training. Would the hens have altered their behavior with an appropriate training regimen? We know from all of the literature on education methods that methods have to be consistent with maturation and intellectual capacity (Vygotsky 1962). Therefore, the fact that recalibration does not occur in some circumstances (e.g., the pecking hens) is not sufficient evidence for concluding that they cannot learn.

The calibrating process can be perhaps best conceived as a behavioral and cognitive background against which recalibration is to be evaluated; a sort of initializing procedure whereby cognitive value is assigned to both sensory input and motor output according to several factors (among which we find the innate perceptual capacity of the organism, its internal states, the task at hand, and the type of stimulus delivered). In this perspective, the calibration process sets cognitive standards by critically correlating sensory input and motor output in given situations. And it is against these standards or this background (a kind of cognitive 'ground zero') that recalibration will be judged to be successful or failed.

As for the relationship between the *capacity* and the *process* of calibration, it can be summed up by saying that the capacity of the system to calibrate is limited by the organism's phylogenetic makeup and its learning activity, whereas the actual calibrating process establishes cognitive standards or markers — within those limits — as functional demands are satisfied and successful behavior is ensured.

## 5. Nervous system plasticity

One important point stands in the way of any clear distinction between innate perceptual capacity and actual perceptual performance in humans: nervous system plasticity. Indeed, as one of the authors' research has shown for many years now, the plasticity of the organism's nervous system plays a crucial role in the development and operation of both the peripheral and central nervous systems, as evidenced by both sensory substitution and rehabilitation medicine studies (Bach-y-Rita 1995). Difficult as it may be to assess the precise extent to which nervous system plasticity contributes to cognitive development, it is clear that some kind of plasticity underlies and explains a good deal of the perceptual system's ability to recalibrate. Indeed, as the experiments with the TVSS and, specially, with modification lenses show, the adaptive power manifested by the perceptual system in its compensating ability is clearly proof of *some type* of plasticity. Many years ago Gibson offered a valuable insight to understand (multimodal) calibration processes via the study of recalibration and its underlying brain plasticity when he stated: "How such a calibration is accomplished in the brain is a problem. But there are experiments to suggest that a *recalibration* occurs when prolonged abnormal information is imposed on a perceptual system, and this may help us to understand the process" (Gibson 1966: 122).

Furthermore, it is useful to keep in mind that the final 'wiring' of the brain takes place after birth and is governed by early experience during infancy. Perceptual calibration thus seems to be a phenomenon that demonstrates the prominent role of active learning and nervous system plasticity in perception.

By now it should be clear that sensory-substitution and perceptual-modification cases are eloquent illustrations of PR. It should also be clear that PR offers an empirical grip on practical and conceptual aspects of such issues as plasticity and development, making explicit, through comparative studies, that which normal calibrating processes maintain implicit.

## 6. The importance of training

Questions regarding the need for training with a substitute sensory system and perceptual modification devices deserve special attention, as much as the brain mechanisms involved in the response to training. Such systems and devices, as we have seen, offer unique opportunities for the study of perceptual systems and relevant conceptual issues. To evaluate the maximum effects of such systems and devices on perception and on behavior in general, and thus to present the strongest possible case for philosophical positions, efficient training must occur. For example, although well-trained blind persons using tactile vision systems appear to "see", we have been reluctant to suggest that blind users of the device are actually seeing, while others (e.g., Heil 1983; Morgan 1977) have not been so reluctant, claiming that since blind subjects are being given similar information to that which causes the sighted to see and are capable of giving similar responses, one is left with little alternative but to admit that they are seeing (and not merely "seeing"). With further training and daily use by blind persons, the issue may become clearer.

Our studies with blind persons have revealed that training is required for tactile vision substitution; and, to a lesser extent, for tactile touch substitution for persons who have lost hand sensation due to leprosy, but have intact sensation proximally (such as on the forehead or trunk, Bach-y-Rita 1995). As has been mentioned, once trained — and this is remarkable — the vibrotactile or electrotactile stimulus array could be moved from the skin region used during training and moved to other regions of the body (e.g., from the skin of the back to that of the abdomen) without requiring any further training. In addition, the motor control of the input could also be shifted (e.g., the TV camera could be moved from the glasses, where movement is controlled by neck muscles, to the hand, where movement is controlled by arm movements) without any retraining required for accurate perception. Thus it appears that both motor and sensory performance was accomplished by brain mechanisms not strictly limited to a localized region. This is not surprising; it occurs as a general rule. Thus, a visual task can be learned with the left eye while the right eye is covered, and can be performed with the right eye while the left eye is covered.

We were surprised to find, however, that initial experiments with a vestibular substitution system for persons who have lost their sense of balance (e.g., due to an ototoxic reaction to an antibiotic) did not need any training. The system consists of a small accelerometer built into a hard hat, and a display (on an electrotactile array on the tongue) that provides tilt information (Tyler, Danilov & Bach-y-Rita 2003). One possibility is that the display is so simple that it is immediately learned, producing nearly instantaneous recalibration. Another possibility is that the subject had already trained herself to substitute tactile input for the lost vestibular system. Six years after the diagnosis of 100% vestibular loss, she had slowly regained the ability to walk, albeit with wide, stumbling gait and requiring the use of a cane. This partial recovery may have been due to her ability to substitute for the loss by a combination of the enhancement of visual and proprioceptive cues, and by the acquired ability to derive balance information from the tactile inputs to the feet. In that case, the tactile input to the tongue from the artificial balance sensor in the hard hat was easily incorporated into her perceptual mechanisms that had already learned to incorporate the foot contact information for a partial sense of balance. At any rate, this does not concern sensory perception but rather proprioception, making uncertain the extent to which PR applies to this kind of case.

## 7. Technology and cognition

The mentioned cases illustrate how technology can affect and modify cognition. Indeed, they involve a human-machine interface wherein prosthetic mediation and its effects on perception are evident.

In this vein, it is worth mentioning the extended capabilities of technologically-mediated perceptual systems, a fact that runs counter to the common-sense notion that cognition is confined to the limits of our body. Similarly, we often think that our skin is the definitive boundary between us and the rest of the world, and that the cognitive agent and the world are two clearly distinct things. But, again, through the use of appropriate instruments and a corresponding perceptual recalibration and training, it is in fact possible to extend our cognition to what *prima facie* is an arbitrarily complex and unlimited spectrum of phenomena.

Experimentation with the TVSS has demonstrated that the type or location of the man-machine interface is not the critical factor for successful cognitive performance, provided that the interface has discriminatory power and a way for the agent to relate the active control of the information pick-up device with an eventual semantic processing of the information. This means that as long as we systematically correlate sensory inputs with behavioral or semantic 'outputs' (via training/practice), we can have access to an indefinitely large range of information and phenomena – some of which are not naturally accessible. An indication that a successful correlation has been stabilized and is operational is provided by reaching the stage of 'transparency' of the interface (i.e., when our awareness shifts from the sensory inputs to the distal object/property that is presented in meaningful terms). Many examples can be offered to illustrate this extended capacity of our perceptual system: from driving a car and feeling the texture of the road 'with our hands', to operating a gigantic crane with sharp dexterity, to feeling anxious when playing a video-game where one is vicariously represented in a game's character. All this boils down to saying that our senses' natural interfaces with the environment can be displaced (by linearly connecting subsequent technological interfaces) and modified with respect to their natural relationship with the environment (by altering the input-output correlations). Cognition is therefore not confined to the limits of the skin (Hutchins 1995; O'Regan & Nöe 2001); and given appropriate technological devices and training, the cognitive boundary between a perceiving agent and its surrounding environment can shift, depending on the purposes at hand. The cognitive boundary between the agent and the world is, therefore, negotiable. One can only imagine what the future will hold as virtual reality technologies more closely resemble the real world and perhaps have the possibility for representing "other worlds".

## 8. Theoretical issues

Let us now turn to some recurrent theoretical issues in the study of sensory substitution and perceptual modification, whose understanding seems to be critical for epistemology, the philosophy of perception, and cognitive science. These issues are: (1) The dynamic character of spatial cognition; (2) Whether the perceptual system's capacity can be determined independently of the functional demands that the environment imposes on the system itself; and (3) The plastic character of the sensory modalities and their typology.

### 8.1 The dynamic character of spatial cognition

We turn this issue into a claim: Based on the preceding discussion, we maintain that spatial cognition is fundamentally dynamic. This is to be understood in a more radical fashion than the mere idea that cognition is accompanied by bodily movement. Indeed, this claim goes beyond the uncontroversial idea that information pick-up is done through an orderly motion of the sensory receptors, and relates to the very nature of perceptual spaces. And it applies to both the phylogenetic and the ontogenetic levels of perceptual development. Hence, our perception of objects in space and of spatial relations is made possible by the fundamentally dynamic character of the cognitive systems, whether this be implicit in their genealogical history or explicit in their actual operation. That perceptual processes *qua* cognitive processes are always of dynamic character provides support for the idea that perception is inextricably linked to directed action in a constraining environment — something that has been repeatedly said and argued for at different times and in diverse forms (e.g., Bergson 1997; Berthoz 1997; James 1952; Gibson 1979; Held & Hein 1963; Varela et al. 1997).

### 8.2 The role of functional demands in perception

The TVSS and other forms of sensory substitution, together with perceptual-modification experiments (cf. Sections 1 and 3), have demonstrated that the degree of adaptive recalibration achieved by individuals (as measured by the success of their behavioral performance and verbal reports) is directly dependent on the functional demands (tasks and inherent goals) imposed on the perceptual system. This is manifest both in the unavoidable training period that implements the recalibration and in the subsequent practice of the acquired abilities.

As we have seen, functional demands provide the goals that guide the system's operation, which becomes stable and viable through the feedback loop 'perception → goal → action' (becoming thence (re)calibrated). This makes sense not only from the point of view of evolutionary neurobiology, ecological and developmental psychology, and control theory, but also in view of the perceptual system's demonstrated plasticity.[6] Conceptually, this functional emphasis on perception emerges from considering that, just as we cannot (re)calibrate an instrument without assigning to it (at least implicitly) a functional goal within certain parametric variations, we cannot ascertain the perceptual system's capabilities without taking into account the functional demands that the system has actually been satisfying throughout its development. From this angle, the study of perception calls for a renewed pragmatist approach in which the perceptual *functions* — rather than the perceptual objects or the perceptual content — reveal what different cognitive organisms and different modalities share in common (see 8.3). These ideas have a well-established foundation in theoretical perceptual psychology (Gibson 1966, 1979; Neisser 1967; O'Reagan & Nöe 2001; Turvey et al. 1981).

## 8.3 The nature and number of sensory modalities

Again, we start with a claim: The number and type of sensory modalities is revisable. This claim is supported by the fact that, through the appropriate instruments and a corresponding perceptual recalibration, it is in fact possible to modify/augment the capabilities of, and even inaugurate, (novel) sensory modalities. Although not on equal footing with the classical modalities in several respects, the new modalities do comply with functional criteria that establish them as new senses in their own right (e.g., *tactovision* — enabled by the TVSS). If this is accepted, then we must revise the traditional Aristotelian credo regarding the number of specialized perceptual senses humans are endowed with (presumably five). Moreover, if the number of sensory modalities is revisable, then it would also seem that the typology of our senses must be revised as well — including intramodal and intermodal boundaries — making the issue an open, and perhaps indeterminate, matter.

However, contemporary accounts of perceptual operation such as Fodor's (1983) defend a neat separation between perceptual modules, in correspondence with which there presumably are specific types of stimuli or perceptual domains.

But the sensory-substitution experiments suggest that conceiving the sensory modalities in terms of isolated, domain-specific channels is problematic at best. From our perspective, the individual senses should rather be viewed as contingent informative channels subordinated to the overall cognitive performance of the organism in a constraining environment; that is, subordinated to the functional demands and the successful behavior of the agent on both an ontogenetic and a phylogenetic scale. True, each channel (including hybrid ones resulting from sensory substitution) has distinct (anatomical, physiological, etiological, phenomenal) traits, but this does not make the senses any less contingent from a functional viewpoint: Modalities subserve function.

We are deliberately not addressing in this paper the qualitative aspect of perceptual experience which, without denying its cognitive import (for it can be argued that qualia are functional and do provide information to the perceiver, beyond a purely objective account of stimulus energy — see Jackson (1982) for a specific example),[7] we view as secondary for the purposes at hand. If pressed on this issue, we should have to admit that — without reducing sensory qualia to mere epiphenomena — the overall cognitive performance of the individual's perceptual systems rules over the qualitative flavor of each specific modality and that, in this sense too, modalities subserve function.

The concept of PR proves useful again, for it reinforces Gibson's (1966) conception of the senses in terms of perceptual systems (i.e., as a whole and complex system that is functionally-constituted as one piece from beginning to end and which depends on the environment). Extending Gibson's ideas, as long as the overall performance of the individual is not compromised, one or several modalities can be lacking, impaired or modified. Indeed, should one or more modalities fail, PR will be activated and — if the organism is to survive and perform successfully — functional compensation will be accomplished. And, as long as the appropriate training and interfaces are adopted, this will be done even if one is not normally calibrated to, say, see with the skin or hear with the eyes.[8]

Furthermore, one could argue for the amodality of perception and maintain that at least some of the underlying representations of objects and concepts are amodal or abstract (Bahrick 2000; Hernandez-Reif & Bahrick 2001; Theios & Amrhein 1989). This idea is especially important in perceptual learning, given the vast amount of information an infant must learn about the environment. Amodal relations simplify reality and provide meaningful linkages between the different senses, as when an approaching vehicle increases in visual angle and sound amplitude (Bahrick 2000). In many respects the important information or meaning in a stimulus is not tied to its specific mode of presentation. For example, correlation in a scatterplot can be accurately perceived through vision, audition (Flowers, Buhman, & Turnage 1997), and touch (Haase & Kaczmarek, in press).

In addition to the above research on amodal perception, much recent evidence supports the view that in many tasks, information from vision, touch, audition is or can be integrated into a single percept (Ernst & Banks 2002; Jackson 2001). This may not always occur and leaves open the possibility for (perhaps task-dependent) modality-specific percepts (Chainay & Humphreys 2002; Macaluso, Frith, & Driver 2000). Similarly, the function of particular areas of the cortex is 'plastic' and capable of 'multimodal' sensory processing; especially in using touch to substitute for vision (D'Angiulli & Waraich 2002). In general then, percepts evolve (and generally become more complex) over time (Perruchet & Vinter 2002) through associative learning, consistent with the findings of many studies in sensory substitution mentioned above. In addition, as we have seen, PR suggests a view that perception is a more dynamic process than is typically recognized (e.g., by the dominant viewpoint of modality specific, static representations). Stated another way, PR involves constructing percepts "on the fly" (Perruchet & Vinter 2002) as opposed to activating templates in memory. Different sensory systems provide the information for perception, but are not the locus of perception itself.

Lastly, a more provocative connection emerges between PR and a concept generating much interest in neuropsychology and consciousness studies —

synesthesia (blending of the senses). One view of synesthesia describes it as a process whereby certain individuals have the capability of tapping into more primitive sensory states (or 'form constants') than most of us are able to (Cytowic 1995). These primitive states are more basic and less modality-specific than the resulting perception. Thus when they hear a sound, for example, it also produces a visual experience, but one that is correlated very tightly with the particular sound. The associations may be arbitrary, but are usually very consistent within a particular individual (i.e., a certain note will always evoke a specific color).

## 9. Conclusions

We hope to have convinced the reader that research within the contexts of sensory substitution and perceptual modification casts a much-needed light on the analysis of sensory perception, technologically-mediated cognition and cognition in general. In particular, we have endeavored to show through those contexts that the concept of PR is, if not necessary, very useful to understand the nature of perceptual adaptation. In our view, the strength of PR resides in its manifold attributes, namely, in its capacity to (1) capture some critical issues that are at stake in the analysis of perceptual adaptation, (2) relate those issues in a relevant way to the analysis of perception and cognition in general, and (3) lend itself to experimentation so as to test some theoretical bases on which our understanding of perception and cognition rely.

Although basically all the empirical literature on perception and adaptation here reviewed supports our argumentation for PR, there remain a few related issues in need of clarification and further study which nevertheless seem to go beyond the experimental data or the theoretical framework provided here. Among these we find:

- *Training and learning methods:* The re/calibrating disposition observed in human and non-human organisms, as mentioned in Section 4, depends on the learning and training methods employed to prompt and refine said disposition. We know that PR must satisfy real functional demands in order for it to arise and be operational, but this by itself does not guarantee the optimality of its implementation procedures or overall outcome.
- *Sensory-motor vs sensory-'motor' perceptual loop:* Although sensory systems are associated with motor systems for perception, in the absence of motor control over the orientation of the sensory input, the sensory part of a sensory-motor loop can be provided by artificial receptors leading to a brain-machine interface (BMI). As noted in note 4, Bach-y-Rita and Kercel (2003)

propose that the motor component of the sensory-motor coupling can also be replaced by a "virtual" movement. They suggested that it is possible to progress to the point where predictable movement, not observed except for some sign of its initiation, could be imagined and by that means the mental image of movement could substitute for the motor component of the loop.

- *Amodal/modality-specific information and concept-formation:* Even though the theoretical framework provided herein for PR privileges an amodal approach for explaining perceptual performance, there seems to be considerable room for discussion in this respect. Indeed, whether neurophysiological, psychophysical or phenomenological, research in these fields makes clear that the debate on the nature of perceptual information and concept-formation remains open (*cf.* Section 8.3).[9]

## Notes

1. Here it is worth mentioning Werner & Wapner's *sensori-tonic field theory* (Werner & Wapner 1952; Wapner & Werner 1957), one that seems to have anticipated some key elements underlying the concept of perceptual recalibration.

2. This is suggested by experiments with severance of cats' optic nerves and cat-rearing in abnormal environments (Calford et al. 2000; Illig et al. 2000).

3. This explicit goal is usually nested within a less explicit series of goals that are simultaneously subserved: from reaching out for a cookie to ensuring survival of the organism.

4. Bach-y-Rita and Kercel (2003) have noted that, under certain special conditions, the mere observance of movement and possibly even the imagination of movement appears to be sufficient to fulfill the "motor" part of the sensory-motor loop. They further suggested that, due to the much faster information transmission of the skin than the eye, innovative information presentation, such as fast sequencing and time division multiplexing can be used to partially compensate for the relatively small number of tactile stimulus points in the BMI. A practical application of such a system would incorporate humans-in-the-loop for industrial applications. It could result in increased efficiency and humanization of tasks that presently are highly stressful.

5. The issue is complicated by the fact that "wiring" (which requires nerve fibers and synapses) is certainly not the only means of information transmission in the brain, and some degree of learning occurs without it. Even single-cell organisms such as amoeba can be said to recalibrate, and invertebrates have been shown to use nonsynaptic diffusion neurotransmission in that process (Bach-y-Rita 1995). Human learning and recalibration, as well as many other processes, may be mediated in part by similar wireless diffusion mechanisms (ibid.). These are often called volume transmission (VT).

6. Indeed, the plasticity of the perceptual system can only be elicited by functional demands.

7. Furthermore, qualia can be linked to emotional aspects of perception, developed through experience (Bach-y-Rita 2003).

8. One of the authors has proposed to displace the emphasis traditionally given in the philosophy of mind to perceptual *content* and focus to functional *performance* and amodal perception, by means of rehabilitating the Aristotelian-Cartesian notion of *sensus communis*. This would lead to a remarkable conceptual economy for relating environment, behavior and evolution around perception, in a pragmatist perspective (González 1999).

9. See also Barsalou et al. (2003).

## References


Bach-y-Rita, P., Collins, C.C., Saunders, F., White, B., and Scadden, L. 1969. "Vision substitution by tactile image projection". *Nature* 221: 963–964.

Bach-y-Rita, P. 1972. *Brain Mechanisms in Sensory Substitution*. New York: Academic Press.

Bach-y-Rita, P. 1995. *Nonsynaptic Diffusion Neurotransmission and Late Brain Reorganization*. New York: Demos-Vermande.

Bach-y-Rita, P., Kaczmarek, K., Tyler, M., and Garcia-Lara, J. 1998. "Form perception with a 49-point electrotactile stimulus array on the tongue". *Journal of Rehabilitation Research and Development* 35: 427–430.

Bach-y-Rita, P. and Aiello, G. L. 2001. "Brain energetics and evolution". *Behavioral and Brain Sciences* 24: 280.

Bach-y-Rita, P. 2003. "Sensory substitution and qualia". In A. Nöe and E. Thompson (eds), *Vision and Mind*. Cambridge: The MIT Press, 497–513.

Bach-y-Rita, P. and Kercel, S.W. 2003. "Sensory-'motor' coupling by observed and imagined movement". *Intellectica* 35: 287–297.

Bach-y-Rita, P., Tyler, M.E., and Kaczmarek, K.A. In press. "Seeing with the brain". *International Journal of Human-Computer Interaction*.

Bahrick, L.E. 2000. "Increasing specificity in the development of intermodal perception. In D. Muir and A. Slater (eds), *Infant Development: The Essential Readings*. Malden, MA: Blackwell, 119–136

Barsalou, L., Simmons, K., Barbey, A., and Wilson, C. 2003. "Grounding conceptual knowledge in modality-specific systems". *Trends in Cognitive Sciences* 7(2): 84–91.

Bergson, H. 1997 [1939]. *Matière et mémoire*. Paris: Quadrige/PUF.

Berkeley, G. 1910 [1709]. *An Essay Towards a New Theory of Vision*. In *A New Theory of Vision and Other Writings*. Introduction by A.D. Lindsay. London: J.M. Dent and Sons Ltd.

Berthoz, A. 1997. *Le Sens du Mouvement*. Paris: Odile Jacob.

Calford, M.B., Wang, C., Taglianetti, V., Waleszczyk, W.J., Burke, W., and Dreher, B. 2000. "Plasticity in adult cat visual cortex (area 17) following circumscribed monocular lesions of all retinal layers". *Journal of Physiology* 524(2): 587–602.

Chainay, H. and Humphreys, G.W. 2002. "Neuropsychological evidence for a convergent route model for actino". *Cognitive Neuropsychology* 19: 67–93.

Cytowic, R.E. 1995. "Synesthesia: Phenomenology and neuropsychology. A review of current knowledge". *PSYCHE* 2(10). http://psyche.cs.monash.edu.au/v2/psyche-2-10-cytowic.html

D'Angiulli, A. and Waraich, P. 2002. "Enhanced tactile encoding and memory recognition in congenital blindness". *International Journal of Rehabilitation Research* 25: 143–145.

Dolezal, H. 1982. *Living in a World Transformed: Perceptual and Performatory Adaptation to Visual Distortion.* New York: Academic Press.

Dretske, F. 1995. *Naturalizing the Mind.* Cambridge, MA: The MIT Press and CNRS.

Ernst, M.O. and Banks, M.S. 2002. "Humans integrate visual and haptic information in a statistically optimal fashion". *Nature* 415: 429–433.

Ewert, P.H. 1930. "The effect of inverted retinal stimulation upon spatially coordinated behavior". *Journal of Genetic Psychology* 7: 177–363.

Flowers, J.H., Buhman, D.C., and Turnage, K.D. 1997. "Cross-modal equivalence of visual and auditory scatterplots for exploring bivariate data samples". *Human Factors* 39: 341–351.

Fodor, J.A. 1983. *The Modularity of Mind.* Cambridge, MA: The MIT Press.

Gibson, J.J. 1966. *The Senses Considered as Perceptual Systems.* London: George Allen and Unwin.

Gibson, J.J. 1979. *The Ecological Approach to Visual Perception.* Boston: Houghton Mifflin.

González, J.C. 1999, "*Sensus communis*, amodal perception and sensory substitution". In *Actas del XIV Congreso Interamericano de Filosofía.* Puebla: AFM/BUAP.

Gregory, R.L. 1981. *Eye and Brain: The Psychology of Seeing.* New York: World University Library.

Haase, S.J. and Kaczmarek, K.A. In press. "Electrotactile perception of scatterplots on the fingertips and abdomen". *Medical and Biological Engineering and Computing.*

Harris, C.S. 1965. "Perceptual adaptation to inverted, reversed, and displaced Vision". *Psychological Review* 72: 419–444

Held, R. 1965. "Plasticity in sensorimotor systems". *Scientific American* 213(5): 84–94.

Held, R. and Hein, A. 1963. "Movement-produced stimulation in the development of visually guided behavior". *Journal of Comparative and Physiological Psychology* 56: 872–876.

Hernandez-Reif, M. and Bahrick, L.E. 2001. "The development of visual-tactual perception of objects: Amodal relations provide the basis for learning arbitrary relations". *Infancy* 2: 51–72.

Heil, J. 1983. *Perception and Cognition.* Berkeley, CA: University of California Press.

Hutchins, E. 1996. *Cognition in the Wild.* Cambridge, MA: The MIT Press.

Illig, K.R., Danilov, Y.P., Ahmad, A., Kim, C.B., and Spear, P.D. 2000. "Functional plasticity in extrastriate visual cortex following neonatal visual cortex damage and monocular enucleation". *Brain Research* 882(1–2): 241–250.

Jackson, F. 1982. "Epiphenomenal qualia", *Philosophical Quarterly* 32: 127–136.

Jackson, S.R. 2001. "'Action binding': Dynamic interactions between vision and touch". *Trends in Cognitive Sciences* 5: 505–506.

James, W. 1952 [1890]. *The Principles of Psychology.* Chicago: Encyclopedia Britannica.

Köhler, I. 1951. "Über Aufbau und Wandlungen der Wahrnehmungswelt". *Österreichische Akademie der Wissensschaften*, Monograph 227.

Köhler, I. 1964. "The formation and transformation of the perceptual world". *Psychological Issues* 3: 1–173 (Monograph 3).

Kottenhoff, H. 1957. "Situational and personal influences on space perception with experimental spectacles". *Acta Psychologica* 15: 79–97 and 151–161.

Macaluso, E., Frith, C., and Driver, J. 2000. "Selective spatial attention in vision and touch: Unimodal and multimodal mechanisms revealed by PET". *Journal of Neurophysiology* 83: 3062–3075.
McCarty, M.E., Clifton, R.K., Ashmead, D.H., Lee, Ph., and Goubert, N. 2001. "How infants use vision for grasping objects". *Child Development* 72(4): 973–987.
McClelland, J.L., and Rumelhart, D.E. (eds) 1986. *Parallel Distributed Processing: Explorations in the Microstructure of Cognition*. Cambridge, MA: The MIT Press.
McKenzie, B.E., Skouteris, H., Day, R.H., Hartman, B., and Yonas, A. 1993. "Effective action by infants to contact objects by reaching and leaning". *Child Development* 64(2) 415–29.
Morgan, M. J. 1977. *Molyneux's Question*. Cambridge: Cambridge University Press.
Morrongiello, B.A., Rocca, P.T. 1989. "Visual feedback and anticipatory hand orientation during infants' reaching". *Journal of Perceptual and Motor Skills* 69: 787–802.
Neisser, U. 1967. *Cognitive Psychology*. New York: W.H. Freeman.
O'Regan, J.K. and Nöe, A. 2001. "A sensorimotor account of vision and visual consciousness". *Behavioral and Brain Sciences* 24(5): 939–1011.
Perruchet, P., and Vinter, A. 2002. "The self-organizing consciousness". *Behavioral and Brain Sciences* 25(3): 297–388.
Piaget, J. 1952. *The Origins of Intelligence in Children*. New York: International Universities Press.
Proust, J. 2000. "Recalibration et représentation mentale". In P. Livet (ed), *De la perception à l'action*. Paris: Vrin.
Rock, I. 1966. *The Nature of Perceptual Adaptation*. New York: Basic Books.
Rock, I. and Harris, Ch.S. 1967. "Vision and touch". In R.C. Atkinson (ed), *Contemporary Psychology: Readings from Scientific American*. San Francisco: W.H. Freeman, 141–149.
Snyder, F.W. and Pronko, N.H. 1952. *Vision with Spatial Inversion*. Wichita, KS: University of Wichita Press.
Stratton, G.M. 1897. "Vision without inversion of the retinal image". *Psychological Review* 4: 341–360, 463–481.
Theios, J. and Amrhein, P.C. 1989. "Theoretical analysis of the cognitive processing of lexical and pictorial stimuli: Reading, naming, and visual and conceptual comparisons". *Psychological Review* 96(1): 5–24.
Turvey, M.T., Shaw, R.E., Reed, E.S., and Mace, W.M. 1981. "Ecological laws of perceiving and acting: In reply to Fodor and Pylyshyn (1981)". *Cognition* 9: 237–304.
Tyler, M.E., Danilov, Y.P., and Bach-y-Rita, P. 2003. "Closing an open-loop control system: Vestibular substitution through the tongue". *International Journal of Integrative Neuroscience* 2(2): 159–166.
Varela, F.J., Thompson, E., and Rosch, E. 1997. *De cuerpo presente: las ciencias cognitivas y la experiencia humana*. Barcelona: Gedisa.
Vurpillot, E. 1963. "La perception de l'espace". In P. Fraisse and J. Piaget (eds), *Traité de Psychologie Expérimentale: VI, La Perception*. Paris: PUF, 97–176.
Vygotsky, L.S. 1962, *Thought and Language*. Cambridge, MA: The MIT Press.
Wapner, S. and Werner, H. 1957. *Perceptual Development. An Investigation within the Framework of Sensory-Tonic Field Theory*. Worcester, MA: Clark University Press.
Werner, H. and Wapner, S. 1952. "Towards a general theory of perception". *Psychological Review* 59: 324–38.

# Distributed processes, distributed cognizers, and collaborative cognition

Stevan Harnad
Université du Québec à Montréal

Cognition is thinking; it feels like something to think, and only those who can feel can think. There are also things that thinkers can do. We know neither how thinkers can think nor how they are able to do what they can do. We are waiting for cognitive science to discover how. Cognitive science does this by testing hypotheses about what processes can generate what doing ("know-how").This is called the Turing Test. It cannot test whether a process can generate feeling, hence thinking — only whether it can generate doing. The processes that generate thinking and know-how are "distributed" within the heads of thinkers, but not across thinkers' heads. Hence there is no such thing as distributed cognition, only collaborative cognition. Email and the Web have spawned a new form of collaborative cognition that draws upon individual brains' real-time interactive potential in ways that were not possible in oral, written or print interactions.

In our age of "virtual reality", it is useful to remind ourselves now and again that a corporation cannot literally have a head-ache, though its (figurative) "head" (CEO) might. And that even if all n members of the Board of Directors have a head-ache, that's n head-aches, not one distributed head-ache. And that although a head-ache itself may not be localized in one point of my brain, but distributed across many points, the limits of that distributed state are the boundaries of my head, or perhaps my body: the head-ache stops there, and so does cognition. If a mother's head-ache is her three children, then her children get the distributed credit for causing the state, but they are not part of the state. And if the domestic economic situation is a "head-ache", that distributed state is not a cognitive state, though it may be the occasion of a cognitive state within the head of one head of state (or several cognitive states in the individual heads of several heads of state).

The problem, of course, is with the vague and trendy word "cognition" (and its many cognates: *cognize*, *cogitate*, *cogito*, and of course the Hellenic forebears: *gnosis*, *agnosia* and *agnostic*). William of Occam urged us not to be profligate with "entities": *Entia non sunt multiplicanda praeter necessitatem.* But entities are just "things"; and we can presumably name as many things as we can think of. Cognition, at bottom, is, after all, *thinking*.

So there are naturally occurring things (e.g., people and animals) that can *think* (whatever thinking turns out to be). Let's call that "Natural Cognition (NC)". And there are artificial things (e.g., certain machines), that can *do* the kinds of things that naturally thinking things can do; so *maybe* they can think too: "Artificial Cognition (AC)". And then there are *collections* of naturally thinking things — or of naturally thinking things plus artificially thinking things — that can likewise do, collectively, the kinds of things that thinking things can do individually, so maybe they can think, collectively, too: "Distributed Cognition" (DC).

But what about the head-aches, and Descartes, and Occam? Descartes was concerned about what we can be absolutely *sure* about, beyond the shadow of a doubt. He picked out the entity we want to baptize as "thinking": It's *whatever is going on in our heads when we are thinking* that we want to call "thinking": We all know *what* that is, when it's happening; no way to doubt that. But we can't be sure *how* we do it (cognitive science will have to tell us that). Nor can we be sure that others can do it (but if they are sufficiently like us, it's a safe bet that they can think too). And we can't even be sure the thinking — however it works — is actually going on within our heads rather than elsewhere (but there too, it's a safe bet that it's all happening inside our heads).

So it seems that head-aches and cognition cover the same territory: heads ache and heads think, and just as it is self-contradictory to deny that my head is aching (when it's aching), it's self-contradictory to deny that my head is thinking (when it's thinking). How do I *know* (for sure) in both cases? Exactly the same way: it *feels like* something to have a head-ache and it *feels like* something to think. No feeling, no aching; by the same token, no feeling, no thinking.

No feeling, no thinking? What about Freud, and the thoughts being thought by my unconscious mind? Well, we have Occam's authority to forget about extra entities like "unconscious" minds unless we turn out to need them to explain what there is to explain. Let's say the jury's still out on that one, and one mind seems enough so far. But can't one and the same mind have both conscious and unconscious (i.e., felt and unfelt) thoughts?

Let's try that out first on head-aches: Can I have both felt and unfelt head-aches? I'd be inclined to say that, inasmuch as a head-ache is associated with something going wrong in my head (a constricted blood vessel, for example), I might have a constricted blood vessel without feeling a head-ache. But do I want to call that an unfelt head-ache? By the some token, if I feel a head-ache *without* a constricted blood vessel, do I then want to say I don't have a head-ache after all? Surely the answer is No in both cases. There are no unfelt head-aches, and when I feel a head-ache, that's a head-ache no matter what else is or is not going on in my head. That's what I *mean* by a head-ache.

So now what about thinking? Can I be thinking when I don't feel I'm thinking? We've frankly confessed that we don't yet know *how* we think; we're waiting for cognitive science to discover and tell us how. But are we ready to accept that — when we are (say) thinking about nothing in particular — we are in fact thinking that "the cat is on the mat", because (say) a brain scan indicates that the activity normally associated with thinking that particular thought is going on in our brains at this very moment? Surely we would say — as with the constricted blood vessel when I do not have a head-ache: "Maybe that activity is going on in my head right now, but that is not what I am thinking! So call it something else — a brain process, maybe. But it's not what I'm thinking unless I actually feel I'm thinking it (at the time, or once my attention is drawn to it).

Perhaps too many years of Freudian profligacy have made you feel that quibbling about whether or not that brain process is a "thought" is unwarranted. Then try this one: Suppose you *are* thinking "the cat is on the mat", but one of the accompanying brain processes going on at that moment is the activity normally associated with thinking that "the cat is *not* on the mat". Are you still prepared to be the thinker of that unfelt thought, the opposite of the one you feel?

Perhaps the years of Freudian faith in the existence of alter egos co-habiting your head have made you ready to accept even that contradiction (but then I would like to test your faith by drawing your attention to a brain process that says you want to sign over all your earthly property to me, irrespective of what might be your conscious feelings about the matter!). At the very least, our subjective credulity about unconscious alter egos — cohabiting the same head but of a different mind about matters than our own — has its objective limits.

A more reasonable stance (and Occam would approve) is to agree that we know *when* we are thinking, and *what* we are thinking (that thought), just as we know when we are feeling a head-ache and what we are feeling (that head-ache), but we do now know *how* we are thinking, any more than we know what processes underlie and generate a head-ache. And among those unknown processes might be components that *predict* that we may be having *other* thoughts at some later time — just as a vasoconstriction without a head-ache may be predictive of an eventual head-ache, and a process usually associated with thinking that "the cat is *not* on the mat" may be predictive of eventually thinking that the cat is not on the mat, even though I am at the moment still thinking that the cat is on the mat.

So we are waiting for cognitive science to provide the functional explanation of thinking just as we are waiting for neurovascular science to provide the functional explanation of head-aches. But we have no doubt about when we are and are not thinking — and almost as little doubt about what we are and are not thinking — as we have of about when we are and are not having a head-ache. There is plenty of scope for unfelt accompanying or underlying *processes* here; just not for unfelt thoughts (or head-aches).

This brings us back to the three candidate forms of thinking: natural/individual (NC), artificial (AC), and collective/distributed (DC). It is important to stress the "collective" aspect of distributed cognition, because of course there is already a form of "distributed" cognition in the natural/individual case (NC). This is thinking that is taking place within one's own head, but with the associated processes being distributed across the brain in various ways. It is not very useful to speak of this as "distributed *cognition*" (DC) at all (though we will consider a hypothetical variant of it later): It is clearly distributed *processes* that somehow underlie and generate the thinking; some of those processes might have felt correlates, many of them may not. For there is another correlate of thinking, apart from the processes that accompany and generate the thinking, and that is the *doing* (or rather the doing *capacity* and *tendency*) that likewise accompanies thinking.

Here it is useful (and Occam would not disapprove) to admit a near-synonym of thinking — namely, *knowing* — that is even more often used as the Anglo-Saxon cognate term for the Latinate "cognition" than "thinking" is. But "knowing" has a liability. It has a gratuitous bit of certainty about it that over-reaches itself, going beyond Descartes' careful delineation of what is certain and necessarily true from what is just highly probably true. When I feel a head-ache, I *know* I have a head-ache, but I merely *think* I have vasoconstriction. That's fine. Both count as cognizing something. But when I think "the cat is on the mat", I certainly don't *know* the cat is on the mat, yet I'm still cognizing. [Strictly speaking, even when I see that the cat is on the mat, I don't know it for sure, in the way that I know for sure that I have a head-ache or that 2+2=4, or that it looks/feels-as-if the cat is on the mat — but there's no need here to get into the irrelevant philosophical puzzles ("the Gettier problems") about the differences between knowing something and merely thinking something that happens to be true.]

The reason knowing is sometimes a useful stand-in for thinking is that it ties cognition closer to action. There is "knowing-that" — which is very much like "thinking-that", as in thinking/knowing that "the cat is on the mat". And there is "knowing-how", as in knowing how to play chess or tennis. (Know-how has no counterpart when we speak only about thinking.) Skill or know-how is something I *have*, and its "proof" is in *doing* it (not just in thinking I can do it). Now an argument can be made for the fact that know-how is not cognition at all. Know-how may (or may not) be acquired consciously and explicitly; but once one has a bit of know-how, one simply has it; conscious thinking is not necessarily involved in its exercise (though one usually has to be awake and conscious to exercise it).

But if know-how were excluded from cognition because it did not necessarily involve conscious thinking, then we would have to exclude all the other unconscious processes underlying the "how" of thinking itself! So, whereas thinking itself is the necessary and sufficient condition for being a cognitive system, thinking in turn has necessary and sufficient conditions of its own too, and most of those are unconscious processes. The same is true of know-how: it is generated by unconscious processes, just as thinking is. Know-how may or may not be acquired, and if acquired, it may or may not be acquired via thinking (though one almost certainly must be awake and thinking while acquiring it); and know-how may or may not be exercised via thinking (though one almost certainly must be awake and thinking while exercising it). Moreover, just about all thinking (including knowing-that) also has a know-how dimension associated with it. If I think that "the cat is on the mat" is true then I know it follows that the "the cat is not on the mat" is false. Thoughts are not punctuate. They have implications. And the implications are part of the know-how implicit in the thought itself. I know how to reply to (many) questions about the whereabouts of the cat if I think that "the cat is on the mat". And the know-how goes beyond the boundaries of thinking and even talking about what I think: it includes doing things in the world. If I think that "the cat is on the mat", I also know how to go and find the cat!

All this belaboring of the obvious is intended to bring out the close link between thinking capacity and doing capacity (via know-how). Which brings us to the second case of cognition — "artificial cognition" (AC). If there are things other than living creatures (e.g., certain machines) that can *do* the kinds of things that living/thinking things can do, then *maybe* they can think too. Note the "maybe". It is quite natural to turn to machines in order to explain the "how" of cognition. Unlike the know-how of the heart or the lungs, the brain's know-how is unlikely to be discoverable merely from observing what the brain can do and what's going on inside it while it is doing so. That might have been sufficient if all the brain could do was to move (in the gross sense of navigating in space and manipulating objects). But the brain can do a lot of subtler things than just walking around and fiddling with objects: it can perceive, categorize, speak, understand, and think. It is not obvious how the know-how underlying all those capacities can be read off brain structure and function. At the very least, trying to design machines that can also do what brains (of humans and animals) can do is a way of testing theories about how such things can be done *at all*, any which way. In addition, it puts the power of both computation and neural simulation in the hands of the theorist.

So, whether or not machines will ever be able to think — and please remember that by "think" we mean being able to have the feeling, rather like a head-ache, that we have agreed with Descartes to call "thinking" — we in any case need machines in order to study and explain the *how* of thinking, the know-how that normally underlies and accompanies the feeling.

Can machines think (Turing 1950)? The answer depends in part on the rather more arbitrary notion we have of what a machine is, compared to our clear, Cartesian notion of what thinking is. Is a system that is designed and assembled out of bio-molecules by people a machine? What if some of its components are synthetic? All of its components? What if toasters grew on trees? Maybe there is no natural kind corresponding to "machine". Maybe all autonomous, moving/functioning systems, whether man-made or nature-made, are machines, and the only substantive question is: Which kinds of machines can do which kinds of things?".

So we are not guaranteed to be right about machine thinking. There will be no Cartesian certainty there. Maybe machines will never be able to *feel*; maybe they will only be able to *do*. So research on AC is, strictly speaking, research on what sorts of machine processes can successfully generate the kinds of know-how we have — can they *do* the kinds of things that we can do, including, of course, speaking and replying (verbally, as well as responding with other coherent actions) to what is said? That is the methodological idea behind the Turing Test: Thinking is as thinking does. Once a machine can do anything a person can do, and do it in a way that is indistinguishable to any person from the way any other person does it, do we have any better grounds for doubting whether the machine thinks than we have for doubting whether any person other than myself thinks? There is no Cartesian certainty in any case but my own.

As soon as we turn from the "whether" question to the "why" question about thinking, we must have recourse to the know-how of machines in order to test theories about the know-how of our brains. That puts AC on a methodological and epistemic par with NC. We have already agreed, moreover, that natural cognition is "distributed" over various parts of the brain, but that it is more accurate to refer to this as the distributed processes underlying or generating cognition (like the distributed processes underlying or generating a head-ache) rather than as "distributed cognition". No doubt machine processes that can do as NC does will be distributed too, across the machine. Or across several machines? We have admitted that the notion of "machine" is fuzzy. By the same token, is the notion of "one machine" not equally fuzzy (Harnad 2003b)?

We have agreed that the machine must be autonomous; it must be able to do whatever it can do *on its own*, without any help from us, otherwise it is partly *our* capacities that are being exhibited and tested in our joint performance capacity.

But whereas a machine's know-how must be independent of ours, does it need to be independent of the know-how of other machines? We are in the same situation here as in the case of the distributed brain processes inside our heads. In the case of our heads, we are talking about the distributed processes that generate two things: our thinking (1) and our know-how (2).

The thinking (1) is a unitary, felt state, like a head-ache; but the physical processes generating it are distributed — distributed, however, only within my head. It is a logical possibility that the physical processes generating my thought "that Venus is a far-away planet" consist of a widely distributed state that includes my head and Venus and perhaps a lot of other components outside my head, just as it is a logical possibility that the physical processes generating my seaside head-ache consist of a widely distributed state that includes my head and the sun and perhaps a lot of other components outside my head. But the probability of such distributed out-of-body feeling-states is of about the same order as the probability of telekinesis, clairvoyance, or reincarnation — or of the possibility that no one other than myself feels. So let us agree to ignore such far-fetched logical possibilities: The limits of my thinking-states are the limits of the processes going on in my brain.

(A bit later, we will consider hypothetical future out-of-body brain prostheses — analogues of artificial kidney machines that perform some brain functions extra-corporeally, via physical links or telemetry. But first we need to return to machine know-how.)

Do we have any reason, with machines, to assume that artificial cognition cannot be distributed across machines?

First, remember that there is an element of Cartesian certainty about natural cognition that has been replaced by mere Turing probability in the case of artificial cognition. The element that has been replaced is actually the essence of thinking, which is that thinking is a form of feeling (1), but a form of feeling that is also closely associated with a capacity and propensity for doing, i.e., with know-how (2). The Turing Test is based purely on (2), with (1) being taken on faith — faith in the same telepathic "mind-reading" powers that each of us uses every day to detect whether and what other people think (Baron-Cohen et al. 2000; Nichols & Stich 2004; Premack & Woodruff 1978).

So if (i) thinking is as thinking does — and hence (ii) if one machine can do anything and everything a thinking person can do, then it can think — then what if one machine cannot do it all, but two, three or ten can? We allowed that the brain processes generating natural cognition could be distributed across the brain: why can't the machine processes generating artificial cognition be distributed across multiple machines?

We are partly up against the arbitrariness of what we mean by (one) "machine" again. It is easy to individuate and count Cartesian thinkers: Each one has a mind (and brain) of his own. Ask them and they will tell you so.You can take a roll call; and if you are one of those thinkers, you know you are not all or part of any of the other thinkers or vice versa. And "mind-reading" aside, the only thoughts (and the only head-aches) each thinker is privy to are his own; and these all occur within the confines of his head. But how do we individuate machines at all (even setting aside the question of whether they can think)?

A toaster seems well individuated: It's a device that bronzes bread. But that seems to pick out an entity only because we are interested in bronzing bread. If the bread-bronzer were just a component of a more complicated Rube-Goldberg device — that drops bread into the bronzing component, which then pops up and hits a bell that triggers a lead ball to roll down a tube onto a roll of toothpaste, whose contents this squeezes onto a rotating electric brush, triggering it into motion — do we have a toaster here or a toothbrush? And how many "machines" are involved?

Individuating machines is not quite as hopeless as this suggests, however, for there are still two non-arbitrary criteria we can use, one sufficient to individuate machines and the other sufficient to individuate "thinking" machines. First, the machine must be *autonomous*: Given its Input (I) it must generate its Output (O) without any outside help (otherwise its boundaries would be indeterminate and it would not have been individuated). Second, its I/O task should be the same as that of a natural living kind: The Turing Test.

So whatever autonomous system can pass the Turing Test counts as one thinking machine: It has the know-how of the corresponding natural thinker (us). (It should be obvious that if one of the components of this machine were itself a natural thinker, that would be cheating, because the whole purpose of the Turing Test is to explain *how* natural thinkers can think, by designing an artificial thinker using components for which we already know how they work. Using a natural thinker as one of the components would just compound the mystery, and leave the "explanation" ungrounded.)

But apart from not including any unexplained components, the Turing-Test-passing machine is free to be any autonomous system that can successfully pass the Turing Test. This entails a lot of constraints already, for our I/O capacity consists entirely of things that we do in space and time with our bodies. So the candidate would have to be a robot; and since it must be able to do anything and everything we can do, in real time, and indistinguishably from the way we do it (it can't navigate a room by sending out a parallel proxy in all directions, nor can it take a lifetime to make a chess move), it has its work cut out for it. Still, there is no reason that all of its hardware has to be located inside the robot. The autonomous system that passes the Turing Test could be a distributed one, with some of its functions inside the robot, others in a remote control station.

Even in the case of human cognition, the future possibility of remote prostheses is not out of the question. What is not negotiable, however, is the autonomy of the system and the unity of feeling, hence thinking. A real brain with synthetic remote prostheses could in principle have a distributed head-ache, with the feeling state literally taking place both in and outside the head. (Remove or deactivate either component and the head-ache vanishes.) So, by the same token, both natural and artificial cognition could be distributed in this sense: The generating processes — already "distributed" spatially within the brain — could have their spatial distribution widened beyond the confines of the person's or robot's head. Nothing really radical about that. But the natural thinker would still be thinking its own individual thoughts, and feeling its own individual head-aches.

About the robot there is no way to know for sure (without actually *being* the robot) whether it is indeed thinking (rather than merely doing, i.e., exhibiting the know-how that is normally generated along with the thinking in the case of thinkers, but without the thinking, because the robot feels nothing at all, neither thoughts nor head-aches). But if the Turing-Test-passing robot is indeed thinking, hence feeling, then it too will be thinking its own individual thoughts and feeling its own individual head-aches, whether its hardware is distributed remotely or all contained locally.

Let us call these two relatively uncontroversial forms of distributedness the "distributed processing" that generates cognition, rather than "distributed cognition" (DC), which we have reserved for the third putative kind of cognition. And let us summarize what is certain, what is probable, and what is possible.

It is certain that I think. It is highly probable that other people think too. It is highly probable that my thinking occurs only within my own head, but that the processes generating my thinking are distributed within my brain. (It is not even clear how "local" as opposed to distributed the brain processes corresponding to a thought would have to be in order to be "non-distributed"; surely even if a thought were generated by the presence of a single molecule, the molecule itself is distributed in space!) It is highly improbable that anything other than humans and other animals can think today. It is highly probable (for the same reason that it is highly probable that other people think too) that anything that can do anything and everything a person can do (indistinguishably from a person, for a lifetime) can think too. So it is highly probable that a robot that could pass the Turing Test would be able to think. It is possible for the processes generating their thinking in both humans and robots to be distributed more widely than just within their respective heads.

Now what about the possibility of true distributed collective cognition, where there is thought generated by distributed processes, some or all of whose constituents are themselves natural or artificial thinkers?

Let us set aside the trivial, unproblematic cases first.

If two people are talking, that's not DC, that's a conversation; same if they're emailing; same if it's n people. Let's call that "collaborative cognition" (CC).

If a person uses a computer or a database, that's not DC, that's human/machine interaction, computation and consultation; with n people jointly using n computers or databases, it's again CC, not DC.

If n people use n computers to gather or process data, that's not DC, that's human/machine interaction and human/human collaboration, i.e., CC; same if n people jointly write and revise a text, or n texts.

If a robot controlled by n people, or n people plus n computers, passes the Turing Test, that too is human/machine interaction and human/human collaboration. It is neither DC nor AC; just NC plus CC (and the Turing Test has not been passed).

If a robot controlled by n computers passes the Turing Test, that is not DC but AC. If the autonomous system consisting of the robot plus not only has know-how, but it also thinks, then that is distributed processing generating thought.

If n robots that can pass the Turing Test email to one another, use computers, gather and process data and jointly write and revise texts, that too is CC, not DC.

So thus far *nothing* is DC. What would it take to generate genuine DC, rather than merely distributed processes generating NC or AC, or distributed NC and AC cognizers collaborating to produce CC? The head-ache test is the decisive one: If the autonomous system consisting of the NC and AC cognizers (plus any other constituents you may wish to add) somehow becomes the kind of system capable of feeling a head-ache, then it is the kind of system capable of thinking a thought, and its constituents have collectively managed to generate DC.

Things nearly as wondrous have happened. If the (distributed) nonliving components and processes inside living single-celled organisms are analogous to the distributed processes that generate NC (and perhaps eventually AC), then the single multi-cellular organism that is generated by the distributed single cells and other components of which it is composed would be analogous to DC. A living thing constituted out of living things is like a thinking thing constituted out of thinking things. But it is highly improbable.

Not to close on an improbable note: Even if there is no DC, but only CC, wondrous things can still arise from it. We could say that all human civilization and knowledge to date already arises from CC. But with the age of the computer and the Internet, the power and possibilities of CC take a quantum leap. Consider the milestones there have been in cognitive evolution (Harnad 1991; Cangelosi & Harnad 2002):

The first cognitive milestone was the evolution of language, millions of years ago, through organic adaptive change in our brains that allowed human cognizers to communicate and collaborate digitally and symbolically, instead of just instrumentally and through sensorimotor imitation, as other species do. This was the greatest cognitive milestone of all, for with it came not only the full power of language to express, describe and explain just about anything, but implicit in it also (although only to be exploited much, much later in human history) was the power of computation to simulate and model just about anything. Language co-evolved with the power of thinking itself (the "language of thought"), and indeed the speed of conversation and the speed of thought are of roughly the same magnitude, allowing cognizers to interdigitate their thoughts, collaborating synchronously, in real time (local CC). Language also allowed human knowledge to be formulated explicitly and to be passed on by word of mouth (the "oral tradition"). This was a form of serial collaboration and accumulation, with each successive teller elaborating the cumulative record in his own way (distal CC).

Being oral, language provided a lot of scope for real-time colloquy and collaboration, but being dependent on serial hearsay for transmission and preservation, its cumulative record was not altogether reliable. So the next cognitive milestone was the invention of writing. This allowed the fruits of human collaboration and thinking to be faithfully recorded, preserved and transmitted speaker-independently — "off-line", so to speak. The offline, asynchronous, written medium thereby became far more powerful than the online, synchronous oral medium for the dissemination, reliability and permanence of human knowledge; but it lacked much of the real-time interactivity for which language and the speed of thought had co-evolved. Hence writing fell out of phase with the potential speed and power of interactive online thought — although it did at the same time foster the skills of solo offline thought: the written tradition.

Print was the third cognitive milestone, radically extending the reach of the written tradition scribbler — independently, but still out of phase with the full speed, power and interactivity of real-time, interdigitating thought. Cognitive collaboration was still either oral and synchronous (leaving no record, until the advent of real-time audio recording) or written and asynchronous, hence far slower and less interactive. Nor did the type-writer or even the word-processor bridge the temporal gap between parallel and serial cognitive collaboration.

The temporal gap between the conversational speed of interdigitating thought for which our brains are adapted and the much slower tempo of dissemination of written text was finally closed by email and the Internet, the fourth cognitive milestone, "scholarly skywriting" (Harnad 2003a). It is now possible for a text to be written, transmitted and responded to in real time, at almost conversational speed (i.e., the speed of thought). Perhaps just as important, it is possible to *quote/comment* text (by live and active or even long dead authors) and to branch that collaborative interaction instantaneously to many other potential interlocutors, and potentially the whole planet, through email, hypermail, blogs, and web archives.

Now, it was never the strength of the oral tradition to have several people speaking at once. Conversation is optimal when it is serial and one-on-one, or with several interlocutors turn-taking — again serially, but in real time. Moreover, not everyone has (or should have) something to say about everything. So there are no doubt constraints and optima that will emerge with skywriting as the practice develops. But right now, the problem is not an excess or embarrassment of sky-written riches, producing an un-navigable din, but a dearth of online scholarly content and CC: Most of cyberspace is still devoted to trivial pursuit, not to CC.

This will soon change. Skywriting itself is one of its own sure rewards. It was the presence of an audience that inspired the eloquence of the bard, the oracle and the sage in the days of the oral tradition. Writing in the skies, instantly visible to one's peers, is one incentive for scholarly CC. So is the prospect (and provocation) of "creative disagreement" (Harnad 1979, 1990). The likelihood of their texts being seen, scrutinized, criticized, used, applied, and built-upon by their peers inspires scholars both to skywrite and to be careful and rigorous; having their skywritings criticized in turn inspires further iterations of skywriting. Soon shared research-data and joint data-analyses too will become part of the skywriting. This is all CC.

The impact of scholarly writing was already being measured and rewarded in Gutenberg days (by counting journal citations); skywriting offers many new ways of monitoring, measuring, maximizing, evaluating, and rewarding the impact of CC through the analysis of (distributed!) patterns in downloads, citations, co-citations, co-authorships, and even co-text (Brody & Harnad 2005).

All of this is CC. It is the fruit of the collective, interactive know-how of many individual thinkers. If it goes wrong, it will inspire many individual head-aches, not one distributed one. And if it inspires pride, that will be felt by many individual cognizers, not one distributed one.

## References

Baron-Cohen, S., Tager-Flusberg, H., and Cohen, D.J. (eds). 2000. *Understanding Other Minds.* New York: Oxford University Press

Brody, T. and Harnad, S. In press. "Earlier web usage statistics as predictors of later citation impact". JASIST http://eprints.ecs.soton.ac.uk/10713/

Cangelosi, A., Greco, A., and Harnad, S. 2002. "Symbol grounding and the symbolic theft hypothesis". In: A. Cangelosi, and D. Parisi (eds) *Simulating the Evolution of Language*. London: Springer. http://cogprints.org/2132

Harnad, S. 1979. "Creative disagreement". *The Sciences* 19: 18–20. http://eprints.ecs.soton.ac.uk/10852/

Harnad, S. 1990. *Creativity: Method or Magic?* http://cogprints.org/1627/

Harnad, S. 1991. "Post-Gutenberg galaxy: The fourth revolution in the means of production of knowledge". *Public-Access Computer Systems Review* 2(1): 39–53. http://cogprints.org/1580/

Harnad, S. 1995. "Interactive cognition: Exploring the potential of electronic quote/commenting". In B. Gorayska and J.L. Mey (eds), *Cognitive Technology: In Search of a Humane Interface*. New York: Elsevier, 397–414. http://cogprints.org/1599/

Harnad, S. 2003a. "Back to the oral tradition through skywriting at the speed of thought". *Interdisciplines*. http://cogprints.org/3021/

Harnad, S. 2003b. "Can a machine be conscious? How?" *Journal of Consciousness Studies*. http://cogprints.org/2460/

Harnad, S. Forthcoming. "Searle's Chinese room argument". In *Encyclopedia of Philosophy*, Macmillan. http://eprints.ecs.soton.ac.uk/10424/01/chineseroom.html

Nichols, S. and Stich, S. 2004. *Mindreading*. Oxford: Oxford University Press.

Premack, D. and Woodruff, G. 1978. "Does the chimpanzee have a theory of mind?" *Behavioral and Brain Sciences* 4: 515–526.

Turing, A.M. 1950. "Computing machinery and intelligence". *Mind* 49: 433–460. http://cogprints.org/499/

# Robotics, philosophy and the problems of autonomy*

Willem F.G. Haselager
Nijmegen Institute for Cognition and Information (NICI)

Robotics can be seen as a cognitive technology, assisting us in understanding various aspects of autonomy. In this paper I will investigate a difference between the interpretations of autonomy that exist within robotics and philosophy. Based on a brief review of some historical developments I suggest that within robotics a technical interpretation of autonomy arose, related to the independent performance of tasks. This interpretation is far removed from philosophical analyses of autonomy focusing on the capacity to choose goals for oneself. This difference in interpretation precludes a straightforward debate between philosophers and roboticists about the autonomy of artificial and organic creatures. In order to narrow the gap I will identify a third problem of autonomy, related to the issue of what makes one's goals genuinely one's own. I will suggest that it is the body, and the ongoing attempt to maintain its stability, that makes goals belong to the system. This issue could function as a suitable focal point for a debate in which work in robotics can be related to issues in philosophy. Such a debate could contribute to a growing awareness of the way in which our bodies matter to our autonomy.

## 1. Introduction

Developments within Artificial Intelligence (AI) influence the way we conceive ourselves; they help to shape and change our self-image as, amongst others, intelligent and autonomous creatures. The field of robotics consists of a particularly noticeable set of tools for the study of basic features of human and animal cognition and behavior (Pfeifer 2004). In this paper I wish to approach robotics, as a cognitive technology (as described by Dascal 2004), from a philosophical perspective and examine the way it plays, as well as could play, a role in the understanding of ourselves as autonomous agents.

Over the years a growing number of claims have been made within AI regarding the existence of autonomous agents. In fact, autonomous agents appear to be a new research paradigm within AI. Candidates for autonomous agents range from software entities (such as search bots on the web) to sophisticated robots operating on Mars. At first sight, robots appear to be reasonable contenders for autonomy for several reasons. First of all, robots are embodied in the sense that their artificial bodies permit real action in the real world. In the case of evolutionary robotics, generally speaking, simulated robots operate in virtual reality, but even in these cases at least certain bodily characteristics and physical features of objects in the world are central to the simulation. Generally, within robotics the focus is on the interaction between morphology, control system and world. Thus, in robotics the emphasis is more on bodily action, whereas the traditional (purely information processing) AI approaches center more on thinking. The capacity to 'do something' makes robots relevant to the topic of autonomy. Secondly, the robots' behavior is often emergent in the sense that it is not specifically programmed for (Clark 2001), and therefore invokes characterizations as 'surprising', 'original' or even 'unwanted'. Moreover, the behavior of robots can be based on their history of interaction with the environment: they can learn (change their behavior over time as a result of the outcome of earlier actions), instead of blindly repeating fixed behavioral patterns. Thus, robots are not only doing things, but often seem to be doing them 'in their own way'. Finally, in many cases when one is observing robots it is hard to refrain from a certain amount of empathy. One immediately starts wondering about what the robot is doing or trying to achieve. Even if they fail, robots often at least seem to have struggled to achieve a goal.

Obviously, many of the words and phrases used above (like action, emergence, original, learn, and struggled to achieve a goal) are ambiguous in the sense that one could argue that these concepts only apply to robots in a metaphorical way, instead of literally, as we generally assume they do in the case of human beings and animals. For many people, robots' bodies, control systems, possibilities for learning and adaptation, and therefore ultimately their behavior is too much dependent on human programming and design in order to speak of genuine autonomy.

A complicating factor in addressing this issue of metaphorical versus literal use of the concept of autonomy is that the meaning of 'autonomy' and 'agency' is far from clear. Like so many other concepts that are central to our self-understanding, the notion of autonomy is difficult to define. Rather it seems to have a variety of meanings, which at best bear a family resemblance to one another, and apply to a great diversity of cases. Similarly, there does not seem to be an absolute cut-off point between autonomous and non-autonomous behavior, but rather a fuzzy border. The only thing that seems to be clear from the beginning is that too permissive and too restrictive interpretations of autonomy need to be avoided. For instance, an understanding of autonomy that would allow thermostats, or one that would only permit adult human beings to qualify, seem inadequate. We may end up with one of them, but to start the investigation on the basis of such extreme interpretations would be question begging.

Adding to the conceptual uncertainty is, as I will try to show below, the difference in interpretations of autonomy between robotics and philosophy. For some this may mean that the debate about autonomy of robots is a non-starter. I would like to suggest however, that it is precisely the interplay between empirical research in robotics and conceptual analysis in philosophy that may help to clarify the confusion. Robotics may profit from a philosophical analysis of the concept of autonomy, because this may lead to a fuller specification of the conditions that need to be fulfilled before robots can be said to have passed the benchmark. On the other hand, philosophy might gain in its attempt to clarify what is (and is not) meant by autonomy, by receiving from robotics concrete and challenging examples of behavior as test cases. Therefore, even though the confusions involved in the issue of the autonomy of robots are considerable, the debate itself can be fruitful for both philosophy and robotics. This debate, in turn, ultimately may deepen our understanding of the important role our bodies play in our autonomy.

## 2. Robots, autonomy and intervention

Historically, robotics grew out of the development of teleoperation (Murphy 2000), i.e., the operating of tools (often more or less like elaborate pliers) from a distance. This way, human operators could avoid having to be present in dangerous circumstances. They manipulated the operator that steered the remote, a tool that functioned for instance in an area with very high temperatures. In the late 1940s teleoperation was used in the first nuclear reactors. These teleoperators were improved by including more feedback to the human operators so they would get a better feeling of what was happening at the remote's end. Manipulating the operators to direct the remote was very tiring and time consuming, and soon the idea arose to have simple and repetitive operations performed automatically, without human steering. This got known as supervisory control: The operator monitors the process and gives high-level commands (e.g., 'turn') while the remote performs parts of the task on its own. All the time the human operators could take over the manipulations. This developed into part-time teleoperation and a semi-autonomous remote. It is no great exaggeration to say that the origins of the roboticist's use of the phrase 'autonomous agents' lie here. Increasing the autonomy of the remote simply means reducing the need for human supervision and intervention. Autonomy is interpreted relative to the amount of on-line (while the robot is operating) involvement of human operators.

Robots go beyond telemanipulators in that their sensorimotor capacities enable them to deal with a greater variety of circumstances and events without on-line human involvement. Robots are considered to be capable to act, in the sense that they not merely undergo events or have effects (like stones rolling down a hill). The robots' operations can be reactive, responding to what is going on (taxes and tropisms), but also proactive, in pursuit of the goals that are active within them, thereby becoming less environmentally driven. Such proactive robots can, as Nolfi and Floreano (2002: 31) put it, be "let free to act", in the sense that they can choose how to achieve these goals.

Fanklin and Graesser (1996) provide a review of the various interpretations of autonomous agents that currently circulate within AI. Rather than repeating that review,[1] I will offer a definition that, I think, captures the general intent: *Autonomous agents operate under all reasonable conditions without recourse to an outside designer, operator or controller while handling unpredictable events in an environment or niche.*

The 'without recourse to an outside designer' refers to recourse *during* the act (i.e., on-line), but not to recourse to a designer preceding the behavior (i.e., programming). Here it is usually pointed out that there is a continuum between complete dependence and complete independence (e.g., Maes 1995: 108). The 'under all reasonable conditions' is added to indicate that there are limits to the robot's functioning ('reasonable'), while at the same time signifying that the robot should not be too dependent on favorable circumstances ('all'). The clause about unpredictable events in an environment is intended to rule out pre-configured systems operating blindly in a completely predetermined environment. Within robotics, then, the increase in autonomy of a system is related to the reduction of on-line supervision and intervention of the operator, programmer or designer in relation to the robot's operations in a changing environment.

## 3. Agents and goals

In the philosophical literature, however, one finds rather more emphasis on the reasons *why* one is acting (i.e., the goals one has chosen to pursue) than on *how* the goals are achieved. Auto-nomos, being or setting a law to oneself, indicates the importance of self-regulation or self-government. Autonomy is deeply connected to the capacity to act on one's own behalf and make one's own *choices*, instead of following goals set by other agents. The importance of being able to select one's own goals is also part and parcel of the common sense interpretation of autonomy.

Although it is impossible to do justice here to all the historical complexities involved, one can trace an opposition between causation through choice versus physical causation back to Plato and Aristotle. In the *Phaedo* (98c–99b), Socrates argues that an explanation of his sitting or lying down purely in terms of his bones, sinews and joints, i.e., in terms of physical or necessary (or, what Aristotle would call efficient) causation, is missing the real cause, namely the reason for his sitting, which is based on Socrates' aim, the result of his choice of what is best to do. Aristotle, in his discussion of the four kinds of causes, emphasized the importance of final causation: "there is the goal or end in view, which animates all the other determinant factors as the best they can attain to; for the attainment of that 'for the sake of which' anything exists or is done is its final and best possible achievement" (*Physics* II, iii; 195a 24–26). For Aristotle, choices are the result of deliberate desires to do something; it is through deliberation that we consider how to put our objectives into practice, and our choices reveal who we are (Hutchinson 1995: 208–210). The essential characteristic of voluntary behavior is that the origin of movement is within the agent's psyche (Juarrero 1999: 16–19).

From this perspective, then, the conception of autonomy within robotics is not very satisfactory. Robots may be operating independently — even 'freely' choosing how to act in order to achieve goals — but the goals they are trying to achieve are still set by human programmers. Although the reduction of on-line involvement of human operators is a considerable achievement for robotics, one could doubt that it is sufficient for the autonomy of the robots involved, because of the enormous amount of human off-line involvement (i.e., the programming and designing of the robot) in advance of the robot's functioning, particularly in relation to which goals the robots should pursue. The contrast between philosophy and robotics regarding this issue is perhaps best illustrated by considering the following set of examples (it is not uncommon to find such cases in philosophical papers on autonomy (see, for instance, Mele 1995; Dennett 2003: 281–285; Mele 2004).

A case of *full* (or harmonious) *autonomy* would be the following: I weigh the pros and cons of drinking beer for a long time, decide that home-made is fine, so I brew my own beer and drink it. This situation is different from a case of *strong will*, where I'd like to drink beer, but since I want to loose weight, I take water instead. This, in turn, differs from the perhaps more familiar case of *weakness of will*; I want to be healthy and think that drinking beer is detrimental to my health, but I buy beer in a shop and drink it all the same. Then there are more extreme cases, such as *delusions*, where I may think that drinking beer is the only way to thwart a plot of aliens to take over the world, so I go to a bar and order a beer. Finally, there is the philosophically popular example of being *brainwashed*. In such a case, external agents make me think that I want to drink beer because it is good for me (or to stop the aliens, or whatever), so I decide to sneak into a beer-factory at night and drink the whole lot.

In these examples, one goes from autonomy of will (selecting a goal) and action (performing a specific behavior) to autonomy of action (weakness of will), to no autonomy at all. Accordingly, the amount of responsibility for my actions decreases (even in the case of delusions I may be considered to bear some responsibility, as I could have taken therapy or medicine, while this option is not available in the case of brainwashing). On the basis of these examples, a philosopher might argue that robots are in a situation comparable to that of people who are brainwashed and that therefore robots are *not even close* to being candidates for any serious degree of autonomy. Robots are not autonomous because they themselves don't choose their own goals and they do not even know it is us that set their goals for them. In relation to the 'true' (philosophical) meaning of autonomy, robots are on the other end of the spectrum.

## 4. Freely choosing goals?

However, that such a conclusion would be a bit premature becomes clear if one considers some developments in the 17th century philosophical debate about autonomy and the ability to choose one's goals. During more or less one century, the clear distinction between choice and necessity started to become quite vague and the notion of final cause fell into disrepute. In relation to the latter, Juarrero (1999: 2) says that purposive, goal-seeking, final causation "no longer even qualified as causal; philosophy restricted its understanding of causality to efficient cause". Efficient causality refers to the push and pull kind of impact of forces on inert matter and it has dominated explanatory practices since the 17th century.

In relation to the former, Vesey (1987: 17) indicates that Descartes introduced a concept of the will that went beyond the classical Greek interpretation of will. According to Vesey, the concept of will was interpreted as 'adopting a favorable attitude to some specific object'. For Descartes, however, 'will' referred to a separate faculty with the power to cause voluntary movements, so that "from the simple fact that we have the desire to take a walk, it follows that our legs move and that we walk (1649: 340). He equated the will with freedom of choice and said: "The will simply consists in our ability to do or not do something" (1641: 40).

However, as soon as volitions are considered to be the causes of actions, Vesey (1987: 20) notes: “it is hard not to allow that volitions are caused by, say, motives, that motives are caused by character, and that character is caused by heredity and environment”. Descartes insisted that the will was free so that the chain of causes would end immediately (“I cannot complain that the will or freedom of choice which I received from God is not sufficiently extensive or perfect, since I know by experience that it is not restricted in any way.” (1641: 39). But Hume went beyond Descartes in claiming that the acts of the will are caused by motives according to the bonds of necessity:

> We may imagine we feel a liberty within ourselves; but a spectator can commonly infer our actions from our motives and character; and even where he cannot, he concludes in general, that he might, were he perfectly acquainted every circumstance of our situation and temper, and the most secret springs of our complexion and disposition” (Hume 1739: 408–409; III, ii).

Not much later, Hartley (1749: 12; Prop. IX) defended a mechanicism, neurophysiologically grounded in his theory of the ‘vibrations of medullary particles’ according to which

> each action results from the previous circumstances of body and mind, in the same manner, and with the same certainty, as other effects do from their mechanical causes; so that a person cannot do indifferently either of the actions A, and its contrary a, while the previous circumstances are the same; but is under an absolute necessity of doing one of them, and that only (Hartley 1749: 84; Conclusion).

Motives are ‘the mechanical causes of actions’ (1749: 86; Conclusion). So, within the century separating Hartley from Descartes, the contrast between events caused by choices versus events caused by necessity had all but disappeared. Although philosophy started out with a conception of agent causation and the capacity to freely choose goals, it ended up in the 18th century with a strikingly mechanistic view. Vesey claims that the natural outcome of this development was formulated most radically in the 20th century by the psychologist Skinner (1953: 447–448) who wrote the infamous *Beyond Freedom and Dignity* (1971) and claimed that the ultimate causes of behavior lie outside the agent and that ‘man is not free’.

## 5. Goals: Having them vs. choosing them

At this point, it would not be unreasonable for roboticists to shrug their shoulders and claim that it is not reasonable to expect a solution to the problem of freedom of will through the development of robots. After all, the issue of freedom of will has turned into one of the major issues in philosophy and it would be far fetched to expect current research in robotics to solve a metaphysical problem that philosophy itself is still trying to come to terms with. As Strawson (1998) says about the problem of free will: "New generations (…) will doubtless continue to launch themselves onto the old metaphysical roundabout". Similarly, as noted earlier, the technical problem of greater independence does not deal with the philosophically interesting aspects of autonomy. These two different versions of the problem of autonomy provide little ground for a debate between robotics and philosophy. In fact, one could even argue that robotics fails as a cognitive technology in relation to autonomy because the gap between what is done and what (philosophically speaking) should be done is too large. Robots, from this perspective, do not provide significant means for the production of useful knowledge concerning autonomy.

However, it is possible to distinguish one more version of the problem of autonomy. This version concerns the question, how and when the goals of creatures genuinely become *theirs*. My goals, at least my basic ones, really belong to me. But when, and on what basis, could we say that robots are pursuing goals of their *own*? This issue of intrinsic *ownership* has to be separated from the harder problem of the freedom of will, i.e., whether we are free to choose what we want to do, and how that would be compatible with physical determinism (and it also has to be distinguished from the technical problem of autonomy within robotics, as described above).

As said, I think it is unrealistic to expect solutions in relation to the problem of free will from robotics. The question concerning the ownership of goals seems to me to lend itself more to input from robotics because, as I will venture below, it raises questions about the *integration* between body, control system, and the aims of the actions undertaken by the system. This is something that can be, and has been, studied both conceptually and empirically.

Even if I don't freely choose my goals, the goals I pursue are *mine*. My goals are important and intrinsically connected to me. Certain goals I do not pursue because others impose them upon me or ask me to achieve them, but because they *matter* to *me*. What makes goals belong to a system? How are they grounded in the system? In the following I will offer a suggestion, that may help to sketch a possible path towards giving an answer to these questions, and I will attempt to show how this can be related to research in robotics.

Fundamentally, what makes my goals mine, is that I myself am at stake in relation to my success or failure in achieving them.[2] That is, goals belong to a system when they arise out of the ongoing attempt, sustained by both the body and the control system, to maintain homeostasis. To a significant extent, it is the body, and its ongoing attempt to maintain its stability, that provides the founding of goals within the system. Autonomy is grounded in the formation of action patterns that result in the self-maintenance of the embodied system and it develops during the embodied interaction of a system with its environment. There are two aspects involved in this suggestion that bear some further investigation in relation to robotics. First of all, there is the issue of the integration between the control system and the body. Secondly, the notion of homeostasis deserves a closer look.

## 6. The integration between body and control system

The biologist von Uexküll (1864–1944) stressed the importance of the integration of all of the organism's components into one purposeful whole. We have to see, he claimed

> in animals not only the mechanical structure, but also the operator, who is built into their organs as we are into our bodies. We no longer regard animals as mere machines, but as subjects whose essential activity consists of perceiving and acting (Uexküll 1957: 6).

Furthermore, he stressed that machines act according to plans of their human designers, but that living organisms are acting plans (Uexküll 1928: 301; see also Ziemke and Sharkey 2001: 708). It is this 'building the operator into the body' that provides, I believe, a profound but legitimate challenge to robotics in relation to the problem of the ownership of goals.

As Chiel and Beer (1997) have pointed out, the brain and the body have developed in constant conjunction during their evolutionary and lifetime interaction with the environment. The search, then, is for an approach that allows for a tight coupling between bodies and control-systems both phylogenetically and ontogenetically. Against this background, the field of evolutionary robotics, with its aim to drive the programmer and designer 'out of the robot' as much as possible, is a very interesting recent development (see, e.g., Sims 1994a 1994b; Nolfi and Floreano 2000).

### 6.1 Evolutionary robotics

Evolutionary robotics has been defined as "the attempt to develop robots and their sensorimotor control systems through an automatic design process involving artificial evolution" (Nolfi 1998: 167). Artificial evolution involves the use of genetic algorithms. The 'genotypes' of robots are represented as bits that can code their morphological features as well as the characteristics (such as weights and connections of a neural network) of their control systems. A fitness formula determines candidates for reproduction by measuring the success of the robots on a specific task. The genotypes of the selected robots are then subjected to crossover with other genotypes and further random mutation, giving rise to a new generation of robots. According to Nolfi (1998: 167–168), the organization of the evolving systems is the result of a self-organizing process, and their behavior emerges out of the interactions with their environment. Therefore, evolutionary robotics is relevant to the topic of autonomy since there is less need for the programmer and/or designer to 'pull the strings' and shackle the autonomy of the evolving creatures, because the development of robots is left to the dynamics of (artificial) evolution.

Artificial evolution is far from straightforward, however, and usually requires an extensive amount of preparation before the evolutionary process can take off. As Nolfi (1998: 179) points out:

> In principle (…) the role of the designer may be limited to the specification of a selection criterion. However, (…) in real experiments the role of the designer is much greater than that: In most of the cases the genotype-to-phenotype mapping is designed by the experimenter; several parameters (e.g., the number of individuals in the population, the mutation and crossover rate, the length of the lifetime of each individual, etc.) are determined by the experimenter; and in some cases the architecture of the controller is also handcrafted. In theory, all these parameters may be subjected to the evolutionary process; however, in practice they are not.

Moreover, it is not unusual to find that the designer not only pre-arranged the evolutionary process, but also interfered directly with its course in order to solve thorny issues such as local minima and the bootstrap problem.[3] In relation to problems such as these, Nolfi (1997) reports, for instance, having to change the fitness formula by adding elements that cause reward for behavior that in itself is not very meaningful, and increasing the number of encounters with relevant stimuli. Often, then, "some additional intervention is needed to canalize the evolutionary process into the right direction" (Nolfi 1997: 196), keeping the designers well in control of their robots, even if the strings to which the robots are tied may be less visible. A second difficulty concerns the fact that, in practice, what evolves is often just the control system and not the body (the morphology of the robot). One of the most used robots is the khepera, a small pre-made robot, with specific sensorimotor capacities that can be controlled through neural networks. In the context of artificial evolution, most often khepera simulators are used instead of the real robots (for reasons of time and costs). The neural networks operate simulated bodies, which are similar in certain respects to the khepera, in a virtual world. After the neural networks have gone through a number of changes during the artificial evolutionary process they can be downloaded into real khepera in a process that could be described as a simple form of 'brain transplantation'. It is important to realize that during the evolutionary process, the khepera itself (both the real one and its simulated counterpart) does not undergo any changes at all. There is no equivalent for this in real evolution. If the integration between body and control system is important for autonomy, it is legitimate to doubt approaches that focus on evolving a neural network in relation to a fixed and pre designed robot body.

However, there is a growing amount of research on the co-evolution of body and control system. Sims (1994a, 1994b) and Harvey, Husbands, and Cliff (1994) provide early examples, and have worked with simulations of robots and environments. More recently, Pollack, Lipson, Hornby, and Funes (2001) have evolved real robots by means of a 3 dimensional printer that uses thermoplastics to build bodies in a flexible way. Here, then, we have body-control system co-evolution and a greater (though by no means complete) emphasis on real (vs. simulated) robots. However, Pollack et al. (2001: 11) note that the types of robots that could be built this way are fairly simple, and conclude:

> The limitations of the work are clearly apparent: these machines do not yet have sensors, and are not really interacting with their environments. Feedback from how robots perform in the real world is not automatically fed-back into the simulations, but require humans to refine the simulations and constraints on design. Finally, there is the question of how complex a simulated system can be, before the errors generated by transfer to reality are overwhelming (Pollack et al. 2001: 14).

Let me summarize. The 'ownership version' of the problem of autonomy leads to the consideration of how goals become grounded in systems. A suggestion I have pursued is that this grounding is based in part on the integration of all bodily components into one purposeful, homeostasis oriented, whole. From this perspective, robotics is interesting because of the developments within evolutionary robotics that aim to model the phylogenetic co-development of the body-control system. Moreover, these developments are accompanied by growing possibilities for translating the simulations into hardware versions, thereby further emphasizing the importance of the interaction between real bodies and real environment. At the same time, however, one has to acknowledge the considerable technical difficulties encountered by recent projects in co-evolutionary robotics. Although it is far too early to draw any clear conclusions, this aspect of the ownership interpretation of autonomy at least creates the possibility for a fruitful debate between philosophers and roboticists about co-evolutionary developed robots.

## 7. Homeostasis: How bodies matter

The notion of homeostasis refers to the regulation of the internal environment of open systems to remain within a region of stability. The concept was introduced by Claude Bernard (1813–1878) and the term by the biologist Walter Cannon in 1932 (meaning: same — steady, i.e., to remain the same). For instance, when glucose concentrations in blood are too high, receptors in the pancreas start a process that results in the release of the hormone insulin, stimulating the conversion of glucose into glycogen that can be stored in the liver, thereby decreasing the glucose concentration. This is simply put, of course, and other examples include oxygen, temperature, water, and urea. This capacity for self-regulation in the service of self-maintenance is characteristic of living organisms. Importantly, homeostasis involves more than just keeping a variable constant through the use of feedback, as in the case of a thermostat regulating temperature, in that the homeostatic system necessarily *depends,* for its own existence, on the self-regulation. A malfunctioning or incorrectly set thermostat need not suffer from the negative consequences it produces, but a truly homeostatic system always will.

Clearly, the notion of homeostasis plays a fundamental role within robotics. Specifically, the relation between homeostasis and self-maintenance is well studied. A simple but perhaps illustrative example concerns the regulation of the amount of energy that a robot has at its disposal. The robot regularly checks its energy level, and takes the necessary actions when the supply runs dangerously low, e.g., by returning on time to its battery reload station. If this kind of self-checking and self-maintaining mechanism does not operate properly, the robot necessarily will suffer from the consequences, as it will simply stop functioning.

However, I would like to suggest that the type of homeostasis that is at issue here is merely functional, but *not genuinely embodied.* To clarify this distinction, let me start by pointing out a well-known difference between robots and living organisms. One can turn robots off for an indefinite amount of time and start them later without any principled problems. A similar procedure is, as we all know, impossible in the case of living creatures. Once 'turned off', they stay off. This is, at least in part, due to the fact that the bodies of current robots are fundamentally different in kind compared to the type of bodies belonging to living organisms. Organic matter decays when not part of a functioning whole, whereas the plastics and metals of robots suffer no such fate.

I think that Maturana and Varela's (1987) concept of *autopoiesis* is particularly relevant to deepen our understanding of this difference. Autopoiesis refers to the self-generating and self-maintaining capacity of the basic building blocks of organic bodies: cells. As Ziemke and Sharkey (2001: 733) say, living organisms consist of autopoietic unities, self-producing and self-maintaining systems. An autopoietic system is a homeostatic machine, and the fundamental variable it aims to maintain constant is *its own organization*. This makes an autopoietic system different from homeostatic machines whose bodies can continue to exist even if they stop operating. The self-organizing capacity of living bodies is based on the autopoietic quality of their basic elements. Such a quality is missing in current robot bodies. Perhaps another way to point to the same difference is to note that an autopoietic system aimed at homeostasis needs to interact continually, for as long as it exists, with its environment. Its basic goals, the ones that really matter to it, *enforce* this continuous interaction (on pains of annihilation). For currently existing robots, the type of homeostasis they are aimed at demands no such thing.

One may think that, again, it would not be unreasonable for roboticists to shrug their shoulders and say that they can hardly be expected to work with organic material and living creatures. There is, after all, a difference between robotics and biology. However, my plea for genuinely embodied homeostasis should *not* be taken as a request to build robots out of living cells, for the notion of autopoiesis does not reflect some intrinsic quality of a specific kind of matter but rather indicates a characteristic of the *organization* of matter. As Maturana and Varela (1987: 51) say: "the phenomena they generate in functioning as autopoietic unities depend on their organization and the way this organization comes about, and not on the physical nature of their components" (see also Ziemke and Sharkey 2001: 732).

Given that it is the organization of the components and not their material constitution that matters, the question is open whether autopoiesis could be realized in artificial matter. All things considered, I do not think that autopoiesis provides a *principled* obstacle for robotics. The argument does indicate a further constraint on robotics, however.[4] Currently robots are constructed mainly out of metals and plastics. A question that needs to be pursued is whether these types of materials allow for a genuinely autopoietic organization. In a way, this brings back Aristotle's ideas about the relationship between matter and form. The form can actualize the potentialities of the matter, but the potentialities have to be there: you cannot build a boat out of sand. A more thorough investigation of the relationship between homeostasis and autopoiesis may lead to the development of what perhaps might be called a *mild* functionalism that pays attention to material aspects of the body to a higher degree than currently is customary. Again, a fruitful debate between philosophers and roboticists concerning this point seems possible.

## 8. Conclusion

Robotics constitutes a valuable cognitive technology because it helps in the understanding of our selves as autonomous agents, also when it is, at times, premature in laying certain claims. Debates about the differences between machines and organisms can further sharpen our knowledge about what actually constitutes autonomy. Analyzing the debate between roboticists and philosophers, I have tried to indicate that they have different conceptions of autonomy, emphasizing, respectively, the capacity for independent (unsupervised) action versus the freedom to choose goals. I have also pointed out some possible historical reasons for this difference. In the course of this paper, I have distinguished three different problems of autonomy. The first one concerns the technical problem of how one selects the right type of behavior to achieve a certain goal. The second problem concerns the hard problem of freedom of will, whether (and if so, how) it is possible to freely choose one's own goals. Finally, there is the issue of how and when goals genuinely belong to the creature itself, instead of merely being imposed upon, or installed within, it. I have suggested that the first problem lacks philosophical import, while the second is out of reach for empirical approaches such as robotics. The third one, however, is relevant for robotics and of considerable philosophical interest as well. Regarding this third problem, one possibility worthy of further investigation is that the capacity to have goals of one's own arises out of the continuous integration of control system and body, resulting in actions aiming at homeostasis. A further understanding of this capacity requires considering the co-evolution of body and control system as well as the specific 'potentialities' of organic matter, i.e., autopoiesis. In relation to both aspects, an exchange between empirical research and conceptual analysis seems possible and potentially fruitful. A collaborative investigation of philosophy and robotics of autonomy might lead to a further strengthening of the growing acknowledgement within cognitive science of the importance of our material constitution for cognition. In relation to our continuing attempt to understand who we are, bodies may matter even more than we currently think.

## Notes

* I would like to thank Susan van den Braak and especially Iris van Rooij for detailed and helpful comments on earlier versions of this paper.

1. Five definitions that are particularly relevant (though not all equally illuminating) here are the following.

(1) Franklin and Graesser (1996: 5): "An autonomous agent is a system situated within and a part of an environment that senses that environment and acts on it, over time, in pursuit of

its own agenda and so as to effect what it senses in the future". (2) Brustoloni (1991,in Franklin 1995: 265): "Autonomous agents are systems capable of autonomous, purposeful action in the real world". (3) Wooldridge and Jennings (1995: 2): "autonomy: agents operate without the direct intervention of humans or others, and have some kind of control over their actions and internal state". (4) Maes (1995: 108): "Autonomous agents are computational systems that inhabit some complex dynamic environment, sense and act autonomously in this environment, and by doing so realize a set of goals or tasks for which they are designed". (5) Murphy (2000: 4): "'Functions autonomously' indicates that the robot can operate, self-contained, under all reasonable conditions without requiring recourse to a human operator. Autonomy means that a robot can adapt to changes in its environment or itself and continue to reach its goal".

2. Of course, there are many 'high-level' goals that are not directly or profoundly related to 'being at stake' (e.g., when I set myself the goal to redecorate the living room). I suggest a similar type of relation between such high-level goals and the more basic ones as between the social (e.g., shame or pride) and more basic (fear or happiness) emotions. That is, in order for the high-level goals to be genuinely mine, the existence of more basic goals that are grounded in my body-control system integration aimed at homeostasis is required.

3. Local minima arise when after a certain amount of progress in relation to the performance of a task, new generations stop improving and the evolving robots get stuck in a sub-optimal performance. The bootstrap problem involves how to get beyond the starting point when the individual robots of the initial generation are unable to differentiate themselves in relation to the task because they all score zero, so that there is no way to select the 'best' or least worst individuals.

4. Pfeifer (2004: 120) provides another reason to emphasize the importance of the materials used for robots: "Most robot arms available today work with rigid materials and electrical motors. Natural arms, by contrast, are built of muscles, tendons, ligaments, and bones, materials that are non-rigid to varying degrees. All these materials have their own intrinsic properties like mass, stiffness, elasticity, viscosity, temporal characteristics, damping, and contraction ratio to mention but a few. These properties are all exploited in interesting ways in natural systems". He argues that robotics similarly should take advantage of the intrinsic properties of matter, in order to simplify the tasks for control systems. Although I am in complete agreement with his argument, my point differs from his in the sense that 'natural matter' not just simplifies the control problem, but also plays a role in grounding the very goals the control system is trying to achieve.

## References

Aristotle. *Physics.* Loeb edition. Cambridge, MA: Harvard University Press.

Brustoloni, J.C. 1991. "Autonomous agents: Characterization and requirements". *Carnegie Mellon Technical Report CMU-CS-91-204.* Pittsburgh: Carnegie Mellon University

Chiel, H.J. and Beer, R.D. 1997. "The brain has a body: Adaptive behavior emerges from interactions of nervous system, body and environment". *Trends in Neurociences* 20(12): 553–557.

Clark, A. 2001. *Mindware: An Introduction to the Philosophy of Cognitive Science.* Oxford: Oxford University Press.

Dascal, M. 2004. "Language as a cognitive technology". In B. Gorayska and J.L. Mey (eds), *Cognition and Technology: Co-existence, Convergence and Co-Evolution*. Amsterdam: John Benjamins, 37–62.
Dennett, D. 2003. *Freedom Evolves*. New York: Viking.
Descartes, R. 1984 [1641]. *Meditations on First Philosophy*. In J. Cottingham, R. Stoothoff, and D. Murdoch (eds), *The Philosophical Writings of Descartes, Vol. 2*. Cambridge: Cambridge University Press, 3–62.
Descartes, R. 1967 [1649]. *Passions of the Soul*. In E.S. Haldane and G.R.T. Ross (eds), *The Philosophical Works of Descartes, Vol. 1*. Cambridge: Cambridge University Press, 330–427.
Franklin, S. 1995. *Artificial Minds*. Cambridge, MA: The MIT Press.
Franklin, S and Graesser, A. 1996. "Is it an agent, or just a program? A taxonomy for autonomous agents". In *Proceedings of the Third International Workshop on Agent Theories, Architectures, and Languages*. Berlin: Springer, 21–35.
Hartley, D. 1970 [1749]. *Observations on Man*. In R. Brown (ed), *Between Hume and Mill: An Anthology of British Philosophy 1749–1843*. New York: The Modern Library, 5–92.
Harvey, I., Husbands, P., and Cliff, D. 1994. "Seeing the light: Artificial evolution, real vision". In D. Cliff, P. Husbands, J.A. Meyer, and S. Wilson (eds), *From Animals to Animats III*. Cambridge, MA: The MIT Press.
Hume, D. 1978 [1739]. *A Treatise of Human Nature*. Oxford: Clarendon Press.
Hutchinson, D.S. 1995. "Ethics". In J. Barnes (ed), *The Cambridge Companion to Aristotle*. Cambridge: Cambridge University Press, 195–232.
Juarrero, A. 1999. *Dynamics in Action: Intentional Behavior as a Complex System*. Cambridge, MA: The MIT Press.
Maes, P. 1995. "Artificial life meets entertainment: Life like autonomous agents". *Communications of the ACM* 38(11): 108–114.
Maturana, H.R. and Varela, F.J. 1987. *The Tree of Knowledge: The Biological Roots of Human Understanding*. Boston: Shambhala.
Mele, A. 1995. *Autonomous Agents*. Oxford: Oxford University Press.
Mele, A. 2004. "Dennett on freedom". *Electronic publication*. Accessed at 29–12–2004. http://gfp.typepad.com/online_papers/files/Al.doc.
Murphy. R.R. 2000. *Introduction to AI Robotics*. Cambridge, MA: The MIT Press.
Nolfi, S. 1997. "Evolving non-trivial behaviors on real robots: A garbage collecting robot". *Robotics and Autonomous Systems* 22: 187–198,
Nolfi, S. 1998. "Evolutionary robotics: Exploiting the full power of self-organization". *Connection Science* 10(3–4): 167–184.
Nolfi, S. and Floreano, D. 2000. *Evolutionary Robotics: The Biology, Intelligence and Technology of Self-Organizing Machines*. Cambridge, MA: The MIT Press.
Nolfi, S. and Floreano, D. 2002. "Synthesis of autonomous robots through evolution". *Trends in Cognitive Sciences* 8(1): 31–37.
Pfeifer, R. 2004. "Robots as cognitive tools". In B. Gorayska and J.L. Mey (eds), *Cognition and Technology: Co-Existence, Convergence and Co-Evolution*. Amsterdam: John Benjamins, 109–126.
Plato. *Phaedo*. Loeb edition. Cambridge, MA: Harvard University Press.
Pollack, J.B., Lipson, H., Hornby, G.S., and Funes, P. 2001. "Three generations of automatically designed robots". *Artificial Life* 7(3): 215–223.
Sims, K. 1994a. "Evolving virtual creatures". *Siggraph '94 Proceedings*, 15–22.

Sims, K. 1994b. "Evolving 3D morphology and behavior by competition". In R.A. Brooks and P. Maes (eds), *Artificial Life IV.* Cambridge, MA: The MIT Press, 28–39.

Skinner, B. 1953. *Science and Human Behavior.* New York: The Free Press.

Skinner, B. 1971. *Beyond Freedom and Dignity.* New York: Bantam Books.

Strawson, G. 1998. "Free will". In E. Craig (ed), *Routledge Encyclopedia of Philosophy.* London: Routledge. Retrieved January 24 2005, from http://www.rep.routledge.com/article/V014SECT5

Uexküll, J. von 1957. "A stroll through the worlds of animals and men: A picture book of invisible worlds". *Semiotica* 89(4): 319–391.

Vesey, G. 1987. "Plato's two kinds of causes". In A. Flew and G. Vesey (eds), *Agency and Necessity.* Oxford: Basil Blackwell.

Wooldridge, M. and Jennings, N.R. 1995. "Agent Theories, architectures, and languages: A survey". In M. Wooldridge and R. Jennings (eds), *Intelligent Agents.* Berlin: Springer, 1–22.

Ziemke, T. and Sharkey, N.E. 2001. "A stroll through the worlds of robots and animals: Applying Jakob van Uexküll's theory of meaning to adaptive robots and artificial life". *Semiotica* 134(1/4): 701–746.

# Technology and the management imagination*

Fred Phillips
Maastricht School of Management

This paper explores the evolution of the techno-management imagination (TMI). This is the process by which, in times of crisis, managers think not just out of the box, but out of the very reality in which the box resides. Tacit social consensus, also known as corporate culture, can lead to a shared, implicit, and incorrect view that certain actions are impossible. TMI transcends local culture, accessing technological solutions that are unknown and/or unimagined. Members of the organization tend to call such solutions "magic". The paper looks at social, perceptual, and managerial aspects of magic from a practical point of view that is grounded in research. It examines the risks of TMI, and concludes with suggested perspectives and research questions for management scientists and cognitive scientists.

## 1. Introduction

This paper explores the evolution of the techno-management imagination (to be abbreviated TMI). This is the process by which, in times of organizational crisis, managers think not just out of the box, but out of the very reality in which the box resides, thus accessing technological solutions that were unknown and/or unimagined. The paper addresses the interaction of technology not just with the artistic and creative imagination as conventionally conceived, but with the perception of ideas and methods that are outside the boundaries of the manager's cultural milieu.

Because these ideas and methods are disallowed by the culture, they are commonly labeled terra incognita, superstition, secret sauce, spirituality, or magic. They are thereby placed in opposition to "knowledge", which is thoroughly embedded in the home culture. One purpose of this paper is to extend scientific discussion toward the treatment of the imaginative unknown.

The discussion addresses cognitive issues of the magical/imaginative and attitudes toward the unknown, and cross-cultural communication, both intra- and inter-organizational. Other cognitive areas to be dealt with in this context are the perception of risk, and, to some extent, terror.

Using examples of technology management (public and private, current and historical), the discussion moves to facilitators and inhibitors of TMI, current issues and possible future directions in TMI, and provides an outline of the high stakes involved in these issues.

It is found that TMI is being exercised more freely and more widely now than in much of the 20th century, though perhaps not as freely as earlier in history. This is occasioned by rapid changes in the socio-technical environment, especially information and communications technology (ICT), which enable decentralized, alliance-based, and entrepreneurial initiatives. The globalization of technology industries is also a factor, as projects become cross-cultural, and affordable ICT democratizes technological initiatives. Managers and scientists should take care to balance the usefulness and urgency of embracing the unknown against the danger of anti-intellectualism and other side effects that can endanger the foundations of technological society. Science can assist this evolving balance by expanding its boundaries to build on newly observed managerial and technological phenomena.

I will argue that technology and magic are now thoroughly interpenetrating, that we therefore need a scientific terminology that will allow us to talk about magic and TMI with intellectual rigor, and that there is both need and justification for managerial excursions into the technological unknown.

After an introduction of background, terminology, and pertinent theory, the paper will look at examples of the ways technology firms are exercising TMI in their operations. It will look at a public-sector foray into magic, the Manhattan Project. Examining some history and philosophy of the interplay of technology and magic to further define a role for TMI in the management of technology, the paper concludes with a discussion of the risks that inhere in modern trends in technology and magic, and an outline of the stakes involved in these risks.

## 2. Magic: Background and basics

Though there is some whimsy in this use of the word "magic", the usage underscores the need to stretch management perceptions as regard the treatment of technology and the search for new technological solutions. It also allows us to make some useful connections to principles of myth. When faced with utterly strange innovations, managers use the word "magic" in tones ranging from awe through whimsy to outright derision. It therefore seems worthwhile to try to sharpen the concept.

This section provides preliminary examples of magic in the context of technology. It musters some relevant theory, using ideas from the philosophy of science, technology life cycle theory, theory of social reality, and cybernetics to provide rigorous vocabulary and workable concepts. It describes the various kinds of socio-technical magic that are the primary focus of this paper. We will see that social magic comes in three flavors: ancient magic, alien (inter-cultural) magic, and innovative magic.

Though this discussion of TMI often refers to "the firm", it should be understood to apply to any organization, public, private, or educational, that invents or uses technology.

## 2.1 Technology life cycle

Current management literature, e.g., Gupta and Smith (2004), distinguishes between exploitation (the extraction of economic value from existing knowledge) and exploration (the search for new opportunities and technologies). Exploration, in this usage, refers to investigation of well-known or readily evident alternatives. That is, the literature does not explicitly address explorations that stretch the very consciousness of the manager.

If it's mainstream, it isn't magic, and so the people in the firm who can access any of the three kinds of magic are a small minority. Figure 1 recalls the familiar characterization of "innovators" as the change-leading minority, and carves out a tiny fraction of the innovators' portion of the bell curve as the domain of wizards, i.e., those who can access magic for the company.

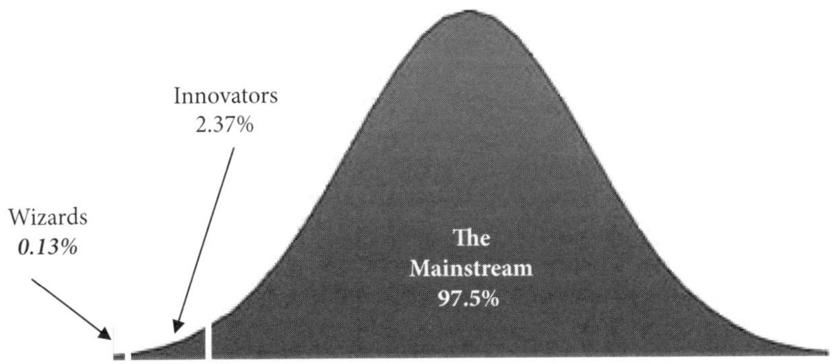


FIGURE 1. Firms in fast-changing markets rely on innovators — that small fraction of the total employee/customer base actively seeking a new product's benefits. Wizards, who are capable of accessing magic (in the sense put forth in this paper), are even rarer than innovators.

Threats to a well-managed firm's very existence should be similarly rare, so we will speak of crises as business situations that are three standard deviations away from the norm. This paper, then, is about matters that are farther from the norm than the problems and solutions managers usually look at. Because they are so rare and seldom attended to, they are effectively the unknown region of the bell curve, the realm of magic. Though rare, they are important because they can respectively threaten the firm's survival and save it. Dealing with them requires a different set of management tools; some are discussed in this paper.

0.13% represents the third standard deviation from the mean of the normal bell curve, a formalism that parallels the usual $2\sigma$ characterization of innovators. (It does not measure the exact proportion of "wizards" in any particular firm.) In a 2004 A.T. Kearney survey (Totty 2004), more than 33% of executives said their companies are either innovators or early adopters of new technologies. This is an increase from the 29% responding similarly in Kearney's 2002 survey, which was already well in excess of the 16% suggested by the classical life cycle theory. It is at best circumstantial evidence that the wizard segment may also be growing, but is encouraging in this regard.

## 2.2 Socially constructed reality

According to an oft-told story, when his exploration vessel first anchored off the reef of Tahiti, Captain Cook watched for a reaction from the natives. There was no reaction. Later conversations revealed that Tahitians on the beach did not see the ship, though it was in plain sight and quite large. How could they fail to see it? The answer has to do with mental categories, denial, and "consensus trance".

Hundreds of years of experience had confirmed that anything on the surface of the ocean was flotsam, a canoe, or a basking whale. Any object not fitting one of those categories could not exist. If a few Tahitians did register something out of the ordinary, they may have been embarrassed to mention it to the majority who showed no signs of noticing. The process by which we actively and tacitly persuade each other to see and believe only a certain subset of the whole is called *consensus trance*, or the *social construction of reality.*

Foreshadowed in the poetry of William Blake, these ideas were first subject to scientific inquiry by Huxley (1963). Later but still seminal works were Berger and Luckmann (1967) and Searle (1997). Berger and Luckmann's book updated Hegel's work on the perception of reality. Searle argues against both materialism (all reality is solely objective) and solipsism (i.e., that reality is solely subjective), in favor of the view that some facts are independent of human observers and some ("intersubjective" facts) require human agreement. Castañeda (1970) put these views into somewhat more accessible language. The cited works imply: (i) Intersubjectivity is essentially what used to be called "mob psychology," and (ii) it applies to a much wider range of experience than crowds and violence; in fact, the intersubjective portion of our daily experience and deepest assumptions is far larger than we might expect.

Does socially constructed reality prevent modern business people from seeing "Cook ships"? Yes, it does. When chaos theory emerged from biology and physics, it became clear that "deterministic chaos" must also be present in market research data. Phillips and Kim (1996) showed how marketing professionals had inadvertently constructed a culture that prevented them from seeing it, despite that taking advantage of these data fluctuations could lead to increased profitability. Other examples appear later in this article.

The following provocative passage supports the thesis that the root of socio-technical magic is a *shared, implicit, and incorrect view that certain things are impossible.*

> Management should never underestimate the constraining effects of technical folklore and deep-rooted prejudices shaped by the accumulated successes and failures of generations of technical products. These attitudes breed barriers so fundamental that they are unrecognizable except to outsiders. Because of such barriers, many companies repeatedly market products that never overcome fundamental flaws and never achieve distinction. Such organizations believe, instinctively and without debate, that certain directions are closed to them, even as other organizations proceed in exactly those directions with success (Rudolph and Lee 1990: 119).

Corporate cultures that repress initiative and lack open lines of communication cannot benefit from the full variety of skills and interests of their constituents. However, what distinguishes TMI from the more ordinary managerial tasks of respecting diversity and leveraging the varied skills and opinions of a diverse employee base (see, e.g., Cummings 2004), are questions of *seeing, denial,* and *teaching.*

- When a "ship of Captain Cook" — something that will change the firm's future irrevocably — appears on the company's horizon, will the company see it? The CIA did not foresee the fall of the Berlin wall. It's not just that they didn't foresee the date; they didn't foresee the *possibility.*

In pursuit of a cohesive company culture, does the company channel employees' perceptions too narrowly, preventing them from seeing Cook ships? S.I. Hayakawa, the linguistics scholar who was president of San Francisco State University, noted: "If you see in any given situation only what everybody else can see, you [are] so much a representative of your culture that you are a victim of it" (http://www.brainyquote.com/quotes/quotes/s/sihayaka153503.html).

- An example and a counter-example:
  - Leaks of sensitive information at the market forecasting firm Forrester Research — due to exploding use of Wi-Fi devices, peer-to-peer file sharing services, instant messaging outside the firewall, and key-fob USB storage devices — led *CIO* magazine to conclude: "The security risks unleashed by rogue technology may far outweigh any productivity gains" (Banham 2004). Yet we must ask, *what if the reverse is true?*
  - 15th century church and city fathers let the contract for the Santa Maria dei Fiori cathedral to an obnoxious clockmaker, Filippo Brunelleschi, making their own reputation by allowing him to realize his vision of an architectural miracle that is still (six hundred years later) the world's largest stone dome (King 2000).
- The CIA did foresee the possibility of the 9/11 tragedy, but — whether due to communication gaps, disbelief, or different priorities — there was a disconnection between the Bush administration and the agency, and hindsight tells us certain actions should have been taken but were not.
- Some Asian languages place little emphasis on the difference between the phonemes "l" and "r". Indeed, Western linguistic science considers them separate phonemes, and Japanese linguistic science (for instance) does not. At the practical level, never bothering to pronounce this difference while growing up, adult speakers of those Asian languages are rarely able to *hear* the difference between els and ars. They consider it mysterious that Westerners can conversationally distinguish the meanings of "lead" and "read". Developmental pathways that allow us to hear the difference do not mature, if we live in a culture that does not place emphasis on the distinction. Of course, very young Asians exposed to western languages have no difficulty articulating els and ars.

## 2.3 Drivers of science and technology

Science progresses when any of its four components — substantive theory, available data, methodology, and the real problems of industry and society — advances, and may leap to the forefront, by building on the current state of the other three (Learner and Phillips 1993). This leapfrog process (Figure 2) occurs within the "paradigms" framework adduced by Kuhn (1970), which describes periodic fundamental shifts in worldview in the sciences. Thus, methodology can sometimes occupy the cutting edge when theory does not: "Science has traditionally been seen as a driving force for technology, but the inverse process is equally important" (McKelvey 1985).

FIGURE 2. A leapfrog model of scientific and technological progress.

## 2.4 The cybernetic view

At IBM and other technology companies (Muehlhausen 2004), innovative capacity is seen as a matter of personal and organizational flexibility. The remaining needed piece of research vocabulary for our discussion of TMI is due to the systems theorist Ross Ashby (1964), who mathematically structured flexibility in organizational and biological systems. An Appendix to this paper describes Ashby's central concept. Theory-oriented readers will find it helpful for understanding "magic" as the term is used in the present paper.

## 2.5 Three flavors of socio-technical magic

A behavior and its outcome may be viewed as magic in any of three circumstances.

*Intercultural magic.* Isolated human cultures of the past had few shared concepts. Their differing living environments led to different senses of what was important. This, in turn, focused their consciousness in ways that led to seeing certain things and not seeing other things. This kind of focus was a survival mechanism.

Thus, a member of Culture X routinely did things that a member of Culture Y (think Navajo and Hopi, or perhaps European crusaders and Middle-Eastern Arabs) would have thought impossible, or at least incomprehensible.

When Culture X and Culture Y collided, because of expanding settlements or chance encounters of hunting parties, each side called the other's odd behavior, and its results, "magic". Chieftains and shamans wishing to maintain the integrity of their own societies would label the other culture's behavior "evil magic", and forbid their own folk to practice it. Today, we are quick to label this attitude as inimical to innovation. However, the ancient tension between deviant behavior and continuity/conservatism of culture is still with us.

It looks like magic when one culture achieves a behavior/outcome pair that is beyond the ordinary horizons of another culture. A member of a traditional subsistence whaling/sealing culture would, no doubt, find it astonishing that modern Americans teach orcas to fling human swimmers high into the air, and let our children place fish in the "killer whale's" mouth.[1] These behaviors are not only possible, they are fun and fairly safe. We may conclude, therefore, that no influential person in the traditional village (where whales were *of course* either quarry or rival seal predators) ever tried to do these things — or, if they thought of it, had no time to attempt it — and so it never occurred to the other villagers that it could be done at all.

In turn, we are astonished by the ways and the reasons Amazonian shamans utilize snakes, frogs, insects and plants. Enterprising pharmaceutical firms are embracing this magic (delicately renamed "ethno-pharmacology") as a cost-effective alternative to discovering disease-fighting drugs in the laboratory. In highly competitive and regulated pharma markets, such drugs can save companies as well as people.

We are also surprised that traditional Asian physicians can diagnose illnesses by taking a patient's pulse and smelling the patient's breath or urine. Western M.D.s confirm these techniques work for a wide range of ailments, but understand that in industrialized nations where medical school and malpractice insurance are expensive, and automated blood diagnostics cheap, there is little point in training doctors in the traditional Asian methods. Now, however, a U.S. university, exercising excellent TMI, has developed a sensor that "smells" abnormal acetone levels in a patient's breath, diagnosing early-stage diabetes more reliably than most other mechanisms (Shapiro 2004).

Corporate cultures seem susceptible to the same effect. When Daimler and Chrysler merged, for instance, executives in each company viewed the others' practices and asked incredulously, "How can they expect *that* to work?" (Executive incentives were one specific focus of skepticism.) The fact was that in each original company, the questioned practice had worked well and long. In the merged company, it could not, and performance of the merged entity bent under the weight of perceptions of bad magic.

*Ancient magic.* Second, a behavior/outcome pair may constitute social magic when it had been a part of a culture, but is no longer needed for the society's survival. The practice may be maintained, by a few, as a cult, a hobby (amateur rocketry would be pertinent to the example below), a specialized niche market, or for entertainment. Maintaining these practices can require extraordinary skill and dedication. In spite of this, or perhaps because of it, practitioners are socially isolated. Shamans, wizards, and artists maintain the integrity of their work by deliberately becoming loners, or even hermits, in order to isolate themselves from social hypnosis.

Where would ancient magic be useful?

- After a generation of space shuttles, NASA has forgotten how to launch high-payload, high-orbit spacecraft. If this knowledge can be recovered, it will be because of carefully archived notes and the capture, perhaps via expert systems, of retired engineers' recollections.
- Finishing violins in the manner of Stradivarius would be of great artistic and economic value today, but the knowledge is lost.
- The first known man-made underwater tunnel joined the two fortified halves of the ancient city of Babylon, in 2160 BC. Nearly four thousand years would pass before the next known tunnel under a river, a passage to transport carts under the Thames (*Popular Science* 1995).

*Innovative magic.* A third category, "innovative magic", is characterized by very advanced science — often theory untested by experiment and for which a practical application has not yet been imagined — combined with audacious engineering, mobilized to save an organization from extreme crisis or competitive pressure. In 1980, shrinking state budgets and the Bayh-Dole Act created a new environment for U.S. higher education. Universities responded by searching their laboratories for commercializeable advances that had remained under the administration's radar. A few universities reaped substantial license income from this new strategy. The discussion below describes the Manhattan Project and Kodak's recent experience in terms of innovative magic.

## 3. How companies use TMI today

Intel is a leading exponent of the "copy exactly" philosophy, which dictates that the first fab to successfully manufacture a new generation of chips should be copied exactly for subsequent facilities, in order to minimize the chance of large downside variation in wafer yield. Another semiconductor firm reportedly built an initial successful fab that, due to a plumbing error, featured a urinal installed five feet above the floor. This company built its second fab — and deliberately placed a urinal in the same inaccessible spot!

Both companies, under intense pressure to recover huge fab construction costs within a short time window, substituted "copy exactly" for a complete scientific understanding of the new chip's manufacturing process. The companies say "copy exactly" demonstrates prudence; others might call it superstition. Differences in personnel, training, local culture, management style, etc. surely result in more yield variation than the placement of the pissoir. In any case, "copy exactly" exemplifies reaching toward the unknown for needed answers at a critical time in a company's history.[2]

In fact, Intel places very little value on *understanding* the semiconductor process (Yeh and Yeh 2004). More engineers than scientists staff Intel's research labs. Engineers "understand" things, of course, and are systematic in their work. But Intel's research staffing shows the company has greater interest in understanding what works, and how to replicate it, than in developing new theory. Intel's need for speed makes the search for theory uneconomical, at least when the theory may be valid for only one chip generation or less.

Referring to any scientific or professional subject, Einstein said, "If you can't explain it to a ten-year-old, you probably don't really understand it yourself" (http://en.thinkexist.com/quotation/if_you_can-t_explain_it_simply-you_don-t/186838.html). Just as well that Intel places small value on understanding; to speak of "understanding" a microprocessor with billions of transistors, that was itself designed in large part by intelligent software, is almost laughable. It is possible only in terms of its testing protocol: If you put that set of signals into a good chip, these numbers should come out. Bad chips, defined and identified as such when these numbers do not come out, are simply discarded. (It may be possible to know how the processor changes a particular input into a certain output, but no one can maintain a mental model of how it transforms myriads of ensembles of inputs.)

Noting the "fluid sense of perception [and] willingness to tinker with cognitive structures" of educated users of hallucinogens, Intel and other major corporations, according to Kirn (1991), give employees ample advance warning for the urine tests they are required to undergo. The desire for creativity in hypercompetitive conditions, he notes, leads to the high-tech industry's "no-sweat attitude toward chemical recreation".

Kodak is replicating the "search the dustiest labs for commercializable advances" strategy, in response to digital photography's powerful threat to Kodak's historic core business. Kodak reports it has, as a result, "found a breakthrough technology that will change the inkjet printer business" (Arner 2004: 97). The company's digital photography business is growing 36% p.a. and by 2005 half of its sales will stem from digital (including printing), even as its traditional (film) photography related sales are in steep decline (*Business Week* 2004).

A less inspiring example comes from AT&T, whose core business of wired telephony was similarly threatened by cellular technology. In 1984, perhaps displaying psychological denial, AT&T badly under-forecasted the number of cell phones that would be in use in 1994. 1994 found AT&T "desperately trying to get back into the business" (Hall 1997: 950). AT&T didn't find magic, and by 2005 it had been bought out by one of its Baby Bell's offspring, SBC.

Firms access magic/TMI by using:

- *Environmental scanning.* Looking outside the firm for new behaviors and technologies.
- *Alliances.* Giant pharmaceutical companies ally with start-ups and universities for drug discovery. In World War II, the British tapped Polish code-breakers, and the Americans utilized Navajo code-talkers.
- *Wizards.* Intel, Xerox, and other technology firms have Fellows programs. Fellows are allowed wide latitude in their choice of projects, and have no corporate management duties. They enjoy the perks of the wizards of myth: Autonomy, solitude, and recognition without the responsibility of political authority.[3] Wizards are found not only in "wizard farms" such as Fellows programs; they may dwell in unexpected places, either within the firm or without.
- The costs of developing a new airframe are so high that Boeing bets the company on every new design. A modern jumbo jet has three million parts. Boeing's 777 development was speeded by a "fly-through" virtual reality system for assembly simulation — a system invented by a physician who hoped to use the technology for surgical training.
- SAP looks for wizards throughout the firm. IBM and Siemens are committed to looking for innovative behavior in "[internal] research… the venture capital community… universities, research institutes, other companies… and the whole environment" (Muehlhausen 2004: M8).

We lose magic when our circumstances evolve. Indeed, hindsight makes what was ordinary before the change magical and legendary after the change. When do we *need* magic? *When circumstances change radically again,* as they are sure to do. Then, old behaviors no longer ensure survival. We have to look to our own old practices, borrow practices from foreign cultures, or invent brand new practices in order to survive.

Using "magic" as a synonym for "something to be resisted", Stuart Cohen, CEO of Open Systems Development Laboratory, remarked: "People thought Linux was magic, that it came mysteriously out of nowhere" (Cohen 2004). When the time comes for magic to save a company, it must have mechanisms in place to help employees see, accept, and perform the magic. Just as the painter Vermeer taught us to *see* the play of morning light through a window, companies have used artists to advance these mechanisms. Shell Oil hired storytellers to turn the company's alternative-futures scenarios into compelling, logical narratives. For the past ten years, Xerox PARC has had a prominent artist-in-residence program.[4] The MIT Media Laboratory has emerged as the leading proponent of artistic participation in the research process. Other firms have actually used stage magicians in their creativity seminars.

## 4. Case: The Manhattan Project

Goethe's poem tells of the Sorcerer's Apprentice (portrayed so ably by Mickey Mouse in *Fantasia)* who prematurely plays with powers beyond his conception. When his master returns, the apprentice cries, "Sir, my need is sore! Spirits that I've excited, my commands ignore".[5]

It took more mature individuals to survive the isolation of the greatest skunk-works project of all time, the Manhattan Project. Robert Oppenheimer was, unquestionably, a wizard. Witnessing the first test of the atomic bomb he had created, he was moved to recite from the Bhagavad Gita. Oppenheimer's familiarity with that profound literature allowed him to reconcile his terrible role in the drama that saved the Allies while it destroyed Hiroshima and Nagasaki.[6] (His momentary identification with the Hindu gods of destruction and regeneration was apposite; under the circumstances, he could be forgiven its grandiosity).[7]

The Manhattan scientists were sure of their science. What made the project *magic* was: (i) Except for the earlier chain reaction achieved under the University of Chicago stadium, it was the first time a suggestion for a practical application of nuclear science had been put into action; (ii) the theoretical possibility of an atomic explosion had never been tested; and (iii) given the untried engineering, the success of the first test explosion bordered on miraculous.

Grilled in 1953 by McCarthy's Un-American Activities Committee, Oppenheimer was not forgiven a pre-war flirtation with communism. His security clearance was revoked and he was removed from his post as Chair of the advisory board for the Atomic Energy Commission. Years later, President Lyndon Johnson politically rehabilitated Oppenheimer at an award ceremony at the White House. Oppenheimer's story illustrates the organizational challenges of dealing with wizards:

- Classically, the wizard vanishes into the mist (or into his lab at the far end of the corporate campus) after the great crisis is resolved — much to the relief of the wizard and everyone else. An ongoing, official, highly public role for Oppenheimer at the AEC would have violated this tradition. A quieter, less official role might have been a great asset to the nation.
- A chief may fear that the successful wizard, now a popular hero, will overshadow him. When it was the chief who recruited the wizard, they may share

credit, and avoid this problem. However, President Roosevelt, who authorized the Manhattan Project, died in 1945. President Truman ordered the use of the bomb on Japan, and the McCarthy witch hunts occurred on President Eisenhower's watch. The House Un-American Activities Committee, with the complicity of Congress, aimed at cutting a number of people down to size, and its deposition of Oppenheimer was consistent with the mythical tradition.[8]

- Self-respecting wizards have no ambition to be mainstream managers, so the fear just mentioned is usually unfounded. When wizards do attempt executive roles, they fail before long, viz., professors and other technology wizards who start companies and must soon relinquish the CEO slot to a professional manager, returning to their R&D or CTO role.
- The "average" wizard is three sigmas away from the rest of us on a number of dimensions. It should hardly have shocked HUAC (House Un-American Activities Committee) that Oppenheimer had other oddities in his past.
- A wizard's impact on society at large is at most intermittent; Oppenheimer's attendance at a meeting or two would hardly impel others toward communism. Artists are more likely to exert a lasting influence on society, and this is one reason HUAC pursued so many in the entertainment industry.

HUAC was an exercise in hysteria. There was, though, no other agency in a position to attempt the good management of Robert Oppenheimer. Had Oppenheimer heeded the lessons of myth, he should have retired to a quiet life of teaching. However, the pull of continued public service was, for someone who could contribute as much as he could, understandably strong.

## 5. Historic interplay of technology and magic

Can there be a philosophical/historical justification for TMI, for technology managers' forays into the unknown? Plato would have said yes; he danced back and forth between rationality and mysticism, and believed the two could not easily be separated. Pascal remarked: "There are but two equally dangerous extremes — to shut reason out, and to let nothing else in" (http://en.thinkexist.com/quotes/blaise_pascal/3.html).

In an excellent modern presentation, Davis (1998) explores the interaction of technological progress and the transcendent imagination.[9] The following elaboration of twelve of Davis' implications ignores, for present purposes, his distinctions between "soul," "spirit" and "magic".

1. *Technologies, real or imagined, are at the very heart of some mythic traditions.* The Masons' view of themselves as the successors of the architect of Solomon's

temple is the perfect example of the Gnostic tradition, equating effective knowledge with the highest pursuit of the spirit. Hermes, messenger of the Greek gods, took a lower road (the Greek *techne* meaning both technique and "trickiness"). That the Masons pursue "hermetic" knowledge illustrates the longstanding link between technology and transcendental imagination.

2. *Technology enables the realization of magical potentials.* Human flight, influence at a distance, strange hybrid animals, etc., are now of course a reality. Modern communications, weaponry, and the World Wide Web give new voice to ancient and obscure beliefs.
3. *Mythical and magical urges help map the possibilities of new technologies.* Example: The "avatars" that represent participants with different levels of privilege and power in computer games and in 2-D and 3-D virtual meeting spaces.
4. *We turn technology to serve the mythic imagination.* Example: Quest games like Myst. Perhaps more extreme are astrological software, I-Ching CD-ROMs, and Tarot hypercard stacks.
5. *Because new technologies are wonder-ful, they make imagination and magic more attractive to the uneducated.* Heron (10–70 AD) built machines (including one that appeared to turn water into wine) the purpose of which was to mystify, impress and baffle. They "paradoxically eroded the cultural authority of the very rational know-how that stimulated their design in the first place" (Davis 1998:19).
6. *Technology can substitute for myth, thus serving as a life jacket for the spiritually sinking,* even if the technology is untested, merely metaphorical, or outright fake. Employing the command-and-control language of the then-new science of cybernetics, L. Ron Hubbard's "Scientology" cult was an attractive beacon for people who felt they had little control over their own lives. Another way of saying this is that mythic yearnings enable behavioral technologies; Davis (1998:126) cites "propaganda, advertising, and mass media, those modern machineries of perceptual manipulation that often explicitly deploy the rhetoric of enchantment".
7. *By the same token, wonderful new technologies force us positively to reframe our myths.* Learner and Phillips (1993:15) described how Hooke's microscope opened new vistas of scientific data, e.g., micro-organisms. Davis (1998:74) adds that these vistas "evoked a sense of wonder and mystery, forcing us to reconfigure the limits of ourselves and to shape the human meaning, if any, of the new cosmological spaces we found ourselves reflected in".
8. Moreover, *each major new basic technology offers new language and metaphors that enhance our persistent myths — and thereby prescribe applications for the technology.* The invention of writing enabled the "religions of the book", Judaism,

Islam and Christianity — and methods of preserving the holy words. Electricity let us "sing the body electric". Reconceptualizing the soul as an electric field, medicine progressed from rigorous galvanic experiments on frogs to bizarre attempts to measure and influence electric phenomena in the human body.

9. *Unfamiliar technology colonizes the mythic imagination.* It may do this in the form of demons, etc. Alien abduction hallucinations constitute an example; they are anxiety dreams about our own mutation. Davis (1998:238) quotes cybertheorist Michael Heim (1997:144): "We experience our full technological selves as alien visitors, as threatening beings who are mutants of ourselves and who are immersed and transformed by technology…".
10. Furthermore, *technology colonizes and animates the extensive universe itself.* Not just in the clutter of machines surrounding us. The creation and study of computers leads us to reconceptualize the universe as an information processor. As man-made computers and computer-based simulations become more powerful, "the universal machine becomes a machine that builds universes". Whether a direct human intention or a matter of complexity theory and AI, researchers have put forth compelling evidence that in the 1990s "the networked noosphere began an irreversible process of self-organization" (Both quotations are from Davis 1998: 126 and 297, respectively. See also Laxton 2000).[10] Medical and biotech advances lead us to create Frankenstein monsters, simulacra of life that themselves echo the mythical Golem.
11. *As hyperlinked technological creation grows complex* (temporarily?) beyond what rational-analytical epistemology can comprehend, we fall back on tribal methods of knowing: "cultivation… pacts, lore, and guiding intuitions" (Davis 1998: 333).
12. And thus, *technology is the root of modern terror.* Marshall McLuhan (1962: 32) said that in highly connected societies, terror is the natural mode of existence, because everything affects everything. Early subsistence-economy societies were tightly connected with their environments because there were no surplus resources to cushion the impact of drought or flood. Today, it is technology that makes globalization possible, tying currencies, product distribution and politics together worldwide.

Writing disseminated the religions of the book. Printing boosted the efficiency of dissemination, as Gutenberg knew it would. Religion was the cause and the effect of printing. Rational technology created LSD, which flings the user into a non-rational world and bounces back to better understanding of schizophrenia and new technologies for psychedelic music. Thus, technologies refine, give expression to, and realize our magical imagination — which forces technology in new directions and causes the cycle to continue. While giving full due to utilitarian uses of technology, Davis echoes Heidegger and Ellul in maintaining that our primeval myth-stories and images are the primary driver of our technology choices, and the primary role of technology is to create new ways of knowing and being.

Among Davis' most powerful points is that members of pre-technological cultures perceived themselves as integral parts of an animated world, in which each stone and tree harbored its own benevolent, mischievous, or malicious spirit. Nanotechnology and advanced electronics will once again complete the interpenetration of the magical and the technical, as today we move toward a techno-animated world where, as in a Disney cartoon, every teakettle will dance on mechatronic legs, sense with silicon/DNA circuits, and speak in a synthesized voice. We can already buy Internet-connected smart refrigerators that monitor milk and order orange juice.

After a false start with a "sphere of knowledge" metaphor, Churchill (2004: 2) recovers: "That metaphor of a sphere is misleading, as if we could glance about and see the work of our confreres. We are more like miners, burrowing away from our once-shared galleries, down shafts… convergent, parallel, divergent". The mining metaphor, not a bad one, has been used before[11] and supports Davis' thrust in point #11 above. Speech is suited for narrative and negotiation, Churchill says, but the Internet lends itself well to hypertext; it is no surprise, therefore, that we cannot comprehend the universe of online knowledge. Churchill notes that the problem has been evident at least since the 1940s, when Bertrand Russell found himself unable to state his philosophical positions in ordinary English.

Thus there are ample philosophical basis and historical precedents for cautiously exercising the techno-management imagination. Indeed, Hall (1997: 19) cites economic historian Henri Pirenne, whose description of the Middle Ages noted "capitalists… incapable of adapting to the conditions that demand needs hitherto unknown and requiring methods hitherto unused… In their place arise new men, bold, entrepreneurial, who allow themselves audaciously to be driven by the wind".

## 6. Cognitive and personality aspects of TMI

This paper's argument rests on the idea that much of what we call cognition is social in nature. Nonetheless, it is legitimate to ask what goes on "inside the head" of an individual wizard, or a manager who abets wizardry. The earlier sections and other literature provide clues to that puzzle. These are reviewed in this section. They remain clues, however, and little more; this section of the paper is of necessity conjectural.

Let us first deal with the question of language. Because consensus trance is tacit ("not looking at the Cook ships"), it is not primarily mediated by language. Language does maintain the pigeonholes into which we drop our experiences. However, the operational principle this implies, i.e., "no pigeonhole = the experience is not real", is one that is transmitted tacitly from person to person. This is one reason the technology transfer literature (e.g., Gibson and Conceição 2003; see also Delcambre, Phillips, and Weaver 2005) emphasizes that "tech transfer is a body contact sport". Managers practicing TMI watch what people do and hear what people say, noting discrepancies between the two.[12] (Future research may test the idea that TMI practitioners are more likely than other managers to have had foreign-language training, sensitizing them to different cultures' pigeonholes and the spaces between them.) ICT makes language-based information ubiquitous. The wizard aims only to insulate him/herself from the powerful social cues that are *not* transmitted by written or spoken language.[13]

Learning style, ego, creativity, self-esteem, and propensity for risk-taking and entrepreneurship seem relevant to an adult's ability to see and use managerial magic. Because children seem less susceptible to consensus trance, it is reasonable to hypothesize that neoteny is a neural and/or personality trait contributing to that ability. The magician's strong sense of self and/or sense of mission makes him or her comfortable far from the center of the home society, and comfortable as well on the borders of other cultures. The enabling manager is comfortable communicating at the social center and at the near fringes; s/he, or an intermediary, is able to meet the wizard on common ground, due to this flexibility. Other psychological factors deserving research attention in this context are the propensity for denial or repression of experiences that cannot be pigeonholed; the developmental neuropsychology of perception; and the drive to "fully understand" vs. the tendency to use partial information in a pragmatic way.

Deakins and Freel (2003: 13) assemble, from several sources, the key personality characteristics of entrepreneurs. Crijns (2005) presents a similar list. Both are presented in Table 1. According to Crijns, the entrepreneurial personality is more likely to engage in the "discovery phase" (initial discovery, opportunity refinement, and market-making) of the innovation process; the managerial personality takes care of the "exploitation phase" (resource acquisition, coordination of new resources, and integration of new and old resources). Several of these characteristics describe wizards (others, e.g., "aggressive", clearly do not), and several describe the manager who is willing to reach out to the unknown. The entrepreneurial characteristics shared by the manager and the wizard are the basis for their communication and cooperation. The same can be said of their complementary

**Table 1.** Key characteristics of entrepreneurs.

| Deakins and Freel (2003) | Crijns (2005) |
|---|---|
| Proactivity, initiative and assertiveness | Initiative |
| Ability to see and act on opportunities | Strong persuasive powers |
| Commitment to others | Moderate risk taker (intrapreneur) |
| Need for achievement | High risk taker (entrepreneur) |
| Calculated risk-taker | Flexibility |
| High internal locus of control | Creativity |
| Creativity | Independence/autonomy |
| Innovativeness | Problem solving ability |
| Need for autonomy | Need for achievement |
| Tolerance for ambiguity | Imagination |
| Vision | High belief in control of one's own destiny |
| Flexibility | Leadership |
| Deviant or non-conformist | Hard work |
| Perseverance; ability to deal with failure | Integrative |
| High energy | Aggressive |
| Emotional stability | Goal-oriented rather than career-oriented |
| Conceptual ability | |
| Charisma | |

characteristics: The wizard may routinely take physical risks, but the manager must be responsible for economic risks.

Entrepreneur and wizard are not synonyms. Table 1 merely helps suggest a specification of the magic-prone personality. We may further hypothesize that the wizard and the enabling manager have experienced multiple cultures (via travel or past employment), have studied history, are capable of both divergent and convergent thinking, and "trust their peripheral vision", that is, their sensory filter does not reject seemingly tangential inputs. Because they are integrative (as Table 1 shows), these individuals cannot help exploring the relation of seemingly tangential stimuli to whatever subject may be the current focus of attention. The result may be fruitful new connections.

We may look also at research on the creative personality. Schiebel (1999) mentions neoteny as contributory to creativity. He adds:

> Clearly, the ability to think "unconventionally" is a key to the creative process. Studies suggest that later-born children in a family are likely to be less parent-oriented and more frequently left to their own devices. This, in turn, seems to encourage less conventional thinking and more creative ideas than that noted in firstborn or only children. Other qualities found in more creative individuals include greater tolerance for ambiguity, more cognitive mobility (the ability to see things from a number of perspectives) and the capacity to change "sets" or problem-solving strategies rapidly.

The entrepreneurship and creativity literature thus suggests the extreme innovator is more devoted to solving the organization's problems than to ego maintenance, career advancement, or a particular methodology. The magician, further, sees both the big picture and the details, so as better to gauge the potential for socio-technical magic to solve the problem.

IBM and Hewlett-Packard both reached outside their cultures for new presidents. It worked at IBM, but not at H-P (Lavelle 2005). IBM's Lou Gerstner absorbed all details of the firm's operations, and mapped effective overall strategy, according to Lavelle, while Carly Fiorina of H-P, also a marketer by background, never shored up or compensated for her lack of operational knowledge. Commentators remarked on Gerstner's devotion to the company. Fiorina, in contrast, left "the impression that she was as interested in burnishing her own image as she was in turning the company around" and "fell in love with her own strategies," at the expense of pragmatic flexibility (Lavelle 2005: 46).[14] Gerstner's retirement left IBM in good shape. Fiorina was fired in 2005 from an H-P that remains in crisis.

The characteristics approach to entrepreneurship has been criticized (Deakins and Freel 2003) on the basis of the mutability of personal traits over time, measurement problems, and its exclusion of environmental factors. The same cautions should apply to the speculations on cognition in this section.

## 7. Socio-technical magic: Risks and stakes

Inhabiting the social fringe, the magician risks developing amazing skills that are relevant to nothing and no one. He or she must balance the need to hone the chosen skills against the possibility of losing all communication with society. The wizard Thomas Edison was eccentric but tolerated. His contemporary and rival Nikola Tesla, possibly the greater intellect of the two, could not or would not come in from the cold regions of the far social fringe.

Other risky aspects of magic:

- Younger people who are still forming a sense of their character and limitations are also at risk if they attempt the wizard role. The earlier reference to Goethe's Sorcerer's Apprentice should suffice to make this point.
- In Phillips (2001), I described a variety of supra-normal phenomena, and noted that fascination with them can lead to estrangement from one's peers, as well as the other perils of idle distraction. I recommended focusing on the everyday affairs of one's business, training the consciousness (perhaps through meditative practice), and having confidence that one will thereby develop TMI, see when magic is needed and, then, where it may be found.

- This same kind of focused practice can help a manager distinguish between intuition and egotism. Listening to intuition, the manager's crisis decisions may be right more often than not. If on the other hand he or she egotistically shoots from the hip because "If I say it, it's right", then the decision will be wrong as often as right.
- In his pioneering work in mathematical decision theory, Pascal (whom I quoted earlier, and who was also an advocate of meditative practice) essayed an analysis of the utilitarian value of religious faith. This analysis is now universally regarded as fallacious. He was trying prematurely to integrate knowledge from his rational inquiries with knowledge from his mystical-philosophical investigations.

## 7.1 Terror

As we have seen, technology enables the globalization that seeds discontents leading to terrorist acts. Technology provides the weapons that are used in these acts. Apart from the atrocities that make the evening news, we suffer generalized technology anxiety as a manifestation of our fear of change and the unnatural.

Technologies also induce terror more or less directly. Employees fear they cannot learn to operate new kinds of equipment. Citizens fear new technology will force new responsibilities upon them (e.g., there is no excuse for not acting the Samaritan when one has a cell phone or CB radio and witnesses a highway accident). Some Americans still panic at the arrival of a long-distance phone call.

By the same token, we use technology to keep terror at bay. Modern Western societies have cleaved rigidly to rationality and religious codes, in order to deny entry to chaos. (The word and the Word are technologies. So are codes, of course, and also all the logical and statistical tools of modern epistemology.) They outlawed LSD. McVeigh and Bin Laden may be willing embodiments of this psychological terror, but are hardly the originators of it. More mundanely, passengers using cell phones prevented United flight 93 from striking the U.S. Capitol on 9/11.

Terror is not only an external threat to social-religious codes, but also a means of enforcing the codes. Davis (1998: 53) notes the respectable press of the early 20th century praised the new method of execution by electric chair, as a way to "imbue the uneducated masses with a deeper terror".

## 7.2 Stakes

The discussion above has suggested several of the losses that can stem from timidity about magic, from indiscriminate use of TMI, and from insufficient appreciation of the interpenetration of technology and magic. The list below recapitulates these and notes additional ones:

1. Reputations are tarred because visionaries and wizards are not managed wisely.
2. Companies fail due to lack of imagination.
3. Resources are misallocated because we do not understand the socio-technical forces which are in play. One venture capital investor speaks of "magic" technologies, noting that VCs invest in entrepreneurs whose technology the VCs don't understand. Instead, investors (close to the social center) perform due diligence by collecting information from people who live between the social center and the far fringes where the techno-wizard entrepreneurs are found. In this way, they find out more about the personality of the entrepreneur as well as more about the viability of the technology. If the statistics are any indication (only one in sixty VC-reviewed business plans leads to an IPO), this is still a badly understood process.
4. Science and technology *must* leapfrog for either to advance more than incrementally. Novelist Poul Anderson maintained that the Celtic civilization could not have grown to become technological in the 21st-century sense because it did not do science. That is, despite the Celts' great craft, they did not admit the existence of universal invariants and regularities; they could not get past the belief that the whimsical imps of the animated world made the universe practically unpredictable. Hall (1997: 25) similarly notes that the ancient Egyptians "could develop technology, but were content to work by trial and error". As Heron has been accused of destroying the reputation of Roman science, and as the Celts and Egyptians could not become technologically advanced, our civilization may lose the capacity (as the animated world re-asserts itself and as machines themselves create new industrial designs and make important decisions) to distinguish between the realm of human decision, design and investigation on the one hand, and the simple wonder (magic) of our new environment on the other.
5. This last point is evident in declining enrollments in science and engineering programs and in the waning level of public technological initiative, in particular for space exploration, that Ray Bradbury (2004) recently mourned in the *Wall Street Journal.* Perhaps in part because we feel less need than we did in the 1950s to seek new wonders elsewhere, we deliberately risk the highest stakes of all — keeping human civilization confined to one lonely, vulnerable planet.
6. In the same vein, Churchill (2004: 2) asserts that the loss, in a hyperlinked world, of our ability to exchange ideas conversationally, puts at stake "the survival of democratic culture".

## 8. Further comments on the meaning of TMI

An influential *Harvard Business Review* article (Nevens et al. 1990) correctly urged companies to prepare to receive and implement innovation, by building cross-functional skills and communication and by building familiarity with a wide range of technologies. However, the authors' emphasis on goal-setting, benchmarking and performance measurement implied a buttoned-down, systematic approach to innovation that was discomfiting. Not only does the latter view seem at odds with millennia of heroic-mythic tradition (viz., *The Lord of the Rings*), but one suspects that companies, when asked, exaggerate the systematic nature of their innovation processes. After all, it would hardly do to tell stock analysts that the firm relies on wizards and wild cards to compete in the marketplace.

The present discussion of the techno-management imagination recognizes the validity of the heroic-mythic tradition, and accepts its applicability in business, within limits. It draws distinctions among creativity, genius, and the simple willingness to embrace the unknown. TMI involves all three, but has a special role for the latter two, as creativity plays a role across the entire bell curve in the daily operations of the firm as well as in its extremes. It does not deny the great value of employees applying creativity to everyday operations, to say that TMI usually requires even more — perhaps what psychologist Howard Gardner calls Exemplary Creators (according to Hall 1997: 11). And TMI can succeed, in given instances, thanks to the "simple willingness", even if genius is not applied.

By citing the writers of socially-constructed reality, I am not endorsing the nihilist element of the postmodern movement. Rather, I am applauding that literature's integration of sociological and psychological aspects of the phenomenon, and hoping that more such integration will appear. Interestingly (and perhaps finding Kuhn's answer to be less than satisfying), according to Hall (1997: 17) Foucault himself wondered how current knowledge gets lost, becoming ancient magic: "How is it that thought detaches itself from the squares it inhabited before… and allows what less than twenty years before had been posited in the luminous space of understanding to topple down into error, into the realm of fantasy, into non-knowledge?" (Hall 1997: 17).

Advocacy of TMI is not tantamount to encouraging "spirituality in the workplace", although spiritual traditions may be among the keys that allow individual managers to find alternative paths to knowledge.

## 9. Conclusion and recommendations for scientists

There may be many kinds of magic. This paper has dealt with just one kind, social (or "socio-technical") magic. It has not implied that science cannot explain this kind of magic — just that (i) science *has* not explained it, or (ii) the magic lies in extraordinary engineering implementation of sound but very advanced science. It has found precedent for the use of the techno-management imagination in organizations, and justification for such ventures beyond well-established knowledge of cause and effect, and has adumbrated some of the risks involved and their stakes.

Action plans addressing each of these risks would be well beyond the scope of this paper. In general, one hopes for a manager's willingness to court magic when the risks to a firm or to the public good justify it, and for scientific investigation to be able to fold the results into an expanded conventional worldview. As for terror, in time, as the technologies and their markets mature and social change stabilizes, one expects productive societies again to muster surplus resources and devote these resources to insulating their day-to-day lives against chaos and terror.

Environmental scientist Sehdev Kumar (2000) wrote: "Real and unreal are a magician's words. It is the great triumph of science that it has extended our vision of the real beyond our senses a millionfold, a billionfold, and has named it the natural world". For scientists, the lessons of TMI are: Investigate the mechanisms by which knowledge becomes lost, taboo, ultimately rediscovered, re-authorized, and integrated with engineering know-how. Investigate the cognitive basis of extreme innovation, and the chains of communication that link the social center with wizards on the fringe. Accept the tripartite roles of science in systematizing knowledge, in extending human imagination, and in archiving and managing knowledge so that lost and culturally strange knowledge can be more easily accessible.

Eulogizing diplomat George Kennan, Holbrooke (2005: A10) captured a key theme of this paper: "As Kennan's life shows, individual, original thinking by one lonely person can sometimes illuminate and guide us better than all the high-level panels and interagency meetings".

## Notes

* This research was supported in part by the Global Business Center at the University of Washington under a grant from the U.S. Department of Education. The author thanks Jonathan Bernstein, David Learner, Dick Raymond, and LaVonne Reimer for comments on an early draft, and a reviewer for this journal for helpful suggestions.

1. This example is adapted from Salzman (1991).

2. The author thanks Dr. Neil Berglund and Dr. Jamie Rogers for discussions of "copy exactly".

3. Intel Senior Fellows represent the company's most exceptional technical professionals", said Craig Barrett, Intel chief executive officer. "Their contributions have helped Intel maintain a position at the forefront of technical innovation and to continuously deliver cutting edge technologies to the marketplace". The Intel Fellow program began in 1980, and in 2002 the company further recognized "the most senior and influential members by creating the new role of Senior Fellow". Employing a total of 65,000 employees, Intel currently has about 50 Fellows and about a dozen Senior Fellows (Intel 2002).

4. "In the words of [PARC] manager John Seeley Brown, the program serves as 'one of the ways that PARC seeks to maintain itself as an innovator, to keep its ground fertile and to stay relevant to the needs of Xerox' Other Silicon Valley, Japanese and some European private firms have followed suit... more or less in agreement that the traditional model of corporate support for the arts — hands-off, patrician, and marketing-driven — overlooks potentials for core innovation" (Century 1999).

5. http://www.fln.vcu.edu/goethe/zauber_e3.html

6. The thrust of the present article is different from Herman Kahn's *Thinking the Unthinkable* (1962), which focuses on outcomes that are horrible in nature, having mostly to do with the use of nuclear weapons. However, by addressing the extension of management thought into "unthinkable" realms, Kahn's subject overlaps the present discussion of social magic.

7. Nearly all biographies of Oppenheimer (e.g., http://www.sparknotes.com/biography/oppenheimer/section8.rhtml) and accounts of the Manhattan Project mention this incident. Of course, any statement about Oppenheimer's frame of mind at the time of this utterance is either inference, or a journalist's notes based on Oppenheimer's later recall of the event. The quotation is imputed to Oppenheimer in two quotation sites, http://www.brainyquote.com/quotes/quotes/j/jrobertop101189.html and http://en.wikiquote.org/wiki/Robert_Oppenheimer, and in Public Broadcasting Systems' short bio of Oppy at http://www.pbs.org/wgbh/amex/bomb/peopleevents/pandeAMEX65.html. In fact, Scott McLemee's review of two Oppenheimer biographies (*Newsday*, 15 May 2005 http://www.mclemee.com/id146.html) is titled "Destroyer of Worlds". McLemee's review is notable for the way it treats the unappreciated mythical aspect of the Manhattan Project.

8. A useful resource for this kind of reasoning is Campbell (1988). A consultant also goes away when the job is done, but a consultant is not a wizard; a consultant would be hard-pressed to make a living specializing in extremely infrequent "3σ" eventualities. Instead, most industry consultants are repositories of current industry practices. That is, industry consultants are sometimes economical substitutes for alliances. University faculty consultants may more closely resemble wizards.

9. As did Nikolai Gogol in his 1836 short story "The Nose": "At that time everyone's minds were tuned to the extraordinary. Just a short time before, the public had been amused by experiments in magnetism.... Someone said that the nose was presently in Junker's store, and... a huge crowd gathered... A respectable-looking entrepreneur with sideburns, who sold various cookies next to the theater entrance, made excellent, sturdy wooden benches and invited the curious to stand on them at the price of eighty kopeks a head". As for the possibility of magical occurrences, Gogol adds: "Say what you may, but such events do happen — rarely, but they do".

10. A news story (Pruitt 2004) points up the magic-complexity-animation link:

> Offering relief from managing complex, distributed systems, Microsoft Corp. Chief Software Architect Bill Gates took the stage of the IT Forum in Copenhagen today [Nov. 16] to introduce a handful of tools and promises. "The magic of software can eliminate this complexity", Gates said in his opening address, amid a cloud of smoke that lingered from the magic show that preceded him onstage. The initiative relies on… having applications and hardware tell management software their status, for instance. Despite some industry criticism that DSI sounds like smoke and mirrors, Gates said Microsoft is proving it a reality with the introduction this week of new offerings aimed at tackling IT complexity.

11. In particular, "data mining" (see Aaker, Kumar, and Day, 2004: 710–714). When, as is often the case, data mining is performed by automated correlation-searching engines, it reveals useful marketing information (for example) while at the same time violating conventional epistemology, insofar as the latter is constructed on the hypothesis-testing principles of classical statistics. See Delcambre, Phillips, and Weaver (2005).

12. Any astute manager can discern the conversational strategies by which colleagues ostentatiously ignore an off-beat idea.

13. Conversely, conferences and conventions are excellent vehicles for conveying new techniques and the sense of new possibilities, because a new social consensus about these items of tacit knowledge can emerge immediately.

14. My own field of management science is well-known for its pragmatic approach to problem solving, which respects but does not idolize existing theory, methodology, or paradigm. Abraham Charnes, a pioneer in the field, was well aware that the existing paradigm creates the conditions that lead to a management situation being perceived as a "problem", and that literature claiming to bear on an unsolved problem merely proves the limitations of the paradigm. Charnes was famous for asking, "Do you want me to read books, or do you want me to solve the problem?" He was not intentionally paradigm-busting. He aimed to solve the industrial problem in a creative manner, and if he could later generalize the solution as a contribution to scientific knowledge, so much the better. Management scientists lampoon the paradigm-bound approach by quoting an economist who, witnessing a working solution to an industrial problem, huffed, "That's fine in practice, but it would never work in theory". Of course, this is not to say that all (or even many) management scientists are wizards.

15. This seemingly simplistic matrix is useful *inter alia* because of the links it makes possible to statistical information theory and game theory.

## References


Aaker, D.A., Kumar, V., and Day, G.S. 2004. *Marketing Research.* 8th (International) edition. Hoboken, NJ: John Wiley & Sons.

Arner, F. 2004. "No excuse not to succeed". *Business Week*, May 10: 96–98.

Ashby, R. 1964. *An Introduction to Cybernetics.* London: Chapman & Hall.

Banham, R. 2004. "Monsters Inc.". *CFO*, April: 71–74.

Berger, P.L. and Luckmann, T. 1967. *The Social Construction of Reality: A Treatise in the Sociology of Knowledge.* New York: Anchor Books.

Bradbury, R. 2004. "To Mars and back". *Wall Street Journal Europe*, November 19: A10.

*Business Week.* 2004. "Kodak's brighter day". *Business Week*, October 4: 46.

Campbell, J. 1988. *The Power of Myth.* New York: Doubleday.

Castañeda, C. 1970. *The Teachings of Don Juan: A Yaqui Way of Knowledge.* New York: Simon and Schuster.

Century, M. 1999. "Pathways to innovation in digital culture". Centre for Research on Canadian Cultural Industries and Institutions. Montreal: McGill University http://www.music.mcgill.ca/~mcentury/PI/PImain.html.

Churchill, J. 2004. "What's it all about, Lord Russell?". *The Key Reporter* 69(4): 2–14.

Cohen, S. 2004. Keynote address, INNOTECH 2004/Oregon Business & Technology Innovation Conference, Oregon Convention Center, Portland, March 31, 2004.

Crijns, H. 2005. "Does entrepreneurship offer opportunities?" Maastricht Management & Marketing Association Entrepreneurship/Intrapreneurship Congress. Maastricht, May.

Cummings, J. N. 2004. "Work groups, structural diversity, and knowledge sharing in a global organization". *Management Science* 50(3): 352–364.

Davis, E. 1998. *Techgnosis: Myth, Magic and Mysticism in the Age of Information.* New York: Three Rivers Press.

Deakins, D. and Freel, M. 2003. *Entrepreneurship and Small Firms.* London: McGraw-Hill Education.

Delcambre, L., Phillips, F., and Weaver, M. 2005. "Knowledge management: A re-assessment and case". *Knowledge, Technology & Policy* 17 (3): 000–000.

Gibson, D.V. and Conceição, P. 2003. "Incubating and networking technology commercialization centers among emerging, developing, and mature technopoleis worldwide". In L.V. Shavinina (ed), *International Handbook on Innovation.* Amsterdam: Elsevier Science, 739–749.

Gupta, A. and Smith, K.G. 2004. "Special research forum call for papers: Managing exploration and exploitation". *Academy of Management Journal* 47(4): 621–622.

Hall, P. 1998. *Cities and Civilization.* New York: Pantheon.

Heim, M. 1997. *Virtual Realism.* New York: Oxford University Press.

Holbrooke, R. 2005. "George Kennan, R.I.P." *Wall Street Journal*, March 22: A10.

Huxley, A. 1963. *The Doors of Perception.* New York: Harper & Row.

*Intel Corporation.* 2002. "Intel appoints first senior fellows", Dec. 9. http://www.intel.com/pressroom/archive/releases/20021209corp.htm

Kahn, H. 1962. *Thinking the Unthinkable.* New York: Horizon Press.

King, R. 2000. *Brunelleschi's Dome: How a Renaissance Genius Reinvented Architecture.* New York: Walker & Co.

Kirn, W. 1991. "Valley of the nerds". *GQ*, July: 96ff.

Kuhn, T. 1970. *The Structure of Scientific Revolutions.* Chicago: The University of Chicago Press.

Kumar, S. 2000. "Why science alone cannot satisfy the soul". *The Globe and Mail*, Toronto, January 17.

Lavelle, L. 2005. "Three simple rules early ignored". *Business Week*, February 28: 46.

Laxton, R. 2000. "The World Wide Web as neural net: Implications for market-driven web enabling". *Technological Forecasting & Social Change* 64(2–3): 55–70.

Learner, D.B. and Phillips, F.Y. 1993. "Method and progress in management science". *Socio-Economic Planning Sciences* 27(1): 9–24.

McKelvey, J.P. 1985. "Science and technology: The driven and the driver". *Technology Review* 88: 38–74.

McLuhan, M. 1962. *The Gutenberg Galaxy*. Toronto: University of Toronto Press.

Muehlhausen, R. 2004. "Corporate innovation: One part genius, nine parts management". The Siemens Round Table on Innovation, Part IV. *Wall Street Journal Europe*: M8.

Nevens M.T., Summe, G.L., and Uttal, B. 1990. "Commercializing technology: What the best companies do". *Harvard Business Review*, May-June: 154–163.

Phillips, F. 2001. *The Conscious Manager: Zen for Decision Makers*. Beaverton, OR: General Informatics.

Phillips, F. and Kim, N. 1996. "Implications of chaos research for new product forecasting". *Technological Forecasting & Social Change* 53(3): 239–261.

*Popular Science*. 1995. "The facts". *Popular Science*, December: 102.

Pruitt, S. 2004. "Gates touts software 'magic' to cut complexity". *ComputerWorld* http://www.computerworld.com/softwaretopics/software/story/0,10801,97556,00.html?source=NLT_AM&nid=97556.

Rudolph, S.E. and Lee, W.D. 1990. "Lessons from the field". *R&D Magazine*, October: 119.

Salzman, M. 1991. *Laughing Sutra*. New York: Knopf.

Schiebel, A. 1999. "Creativity and the brain". Public Broadcasting Service: http://www.pbs.org/teachersource/scienceline/archives/sept99/sept99.shtm

Searle, J.R. 1997. *The Construction of Social Reality*. New York: Free Press.

Shapiro, S. 2004. "Medicinal weeks and diabetes". *Business Week*, July 5: 71.

Totty, M. 2004. "Payoffs for pioneers". *Wall Street Journal Europe*, November 19: R7.

Yeh, R. and Yeh, S. 2004. *The Art of Business: In the Footsteps of Giants*. Olathe, CO: Zero Time Publishing.

## Appendix: A cybernetic construction of magic

Valuable research vocabulary for a discussion of magic is due to the systems theorist Ross Ashby (1964), who mathematically structured flexibility in organizational and biological systems. This Appendix describes Ashby's central concept.

Table A1. A model of the firm showing environmental stimuli, behavioral responses, and outcomes.

| | | The firm's responses to its environment | | | |
|---|---|---|---|---|---|
| | | α | β | γ | δ |
| | 1 | *b* | *d* | *a* | *a* |
| Environmental | 2 | *a* | *d* | *a* | *d* ← Outcomes |
| States | 3 | *d* | *b* | *a* | *b* |
| | 4 | *d* | *a* | *b* | *d* |

Table A1 depicts the way an organization responds to its environment.[15] In business terms, "environmental states" or stimuli might include such things as: Economic recession, demographic shifts in the national market, a new competitor entering the market, or an old competitor cutting its prices. Examples of a firm's responses (the columns α–δ) include replacing the CEO, restructuring salesforce compensation, changing the advertising mix, introducing more new products, or, really, any business decision. Though the Table shows four "outcomes" a–d, two outcomes are usually of practical concern: The trend in stock price stays in a band acceptable to shareholders, or it does not.

Looking at Table A1, let us assume that outcome "b" is conducive to the continued success of the firm, and the other outcomes are not. The Table shows that this firm has behavioral arrows in its quiver that will allow it to succeed under any of the business environments it normally encounters. For example, if the business climate is "3", the firm responds with behavior "β", and felicitous outcome "b" is achieved.

Now suppose an unexpected environmental factor comes into play. It may be a business condition that was previously thought to be so low-probability that it was uneconomical to maintain a ready response for it; or it might be a completely new-to-the-world environmental condition. A real example of such an unexpected stimulus was the Tylenol contamination incident. We represent the new condition as row 5 of Table A2.

Table A2. A model of the firm showing a new environmental stimulus.

| | | The firm's responses to its environment | | | |
|---|---|---|---|---|---|
| | | α | β | γ | δ |
| | 1 | *b* | *d* | *a* | *a* |
| Environmental | 2 | *a* | *d* | *a* | *d* |
| States | 3 | *d* | *b* | *a* | *b* ← Outcomes |
| | 4 | *d* | *a* | *b* | *d* |
| | 5 | *d* | *a* | *a* | *a* |

It is easy to see that in the scenario of Table A2, none of the established responses assure a favorable "b" outcome. This firm will have to look beyond its established responses — perhaps far beyond — to find a new response (call it ε) that brings the outcome back into the survival range. See Table A3.

**Table A3.** A model of the firm showing an effective new response to a new stimulus.

| | | The firm's responses to its environment | | | | |
|---|---|---|---|---|---|---|
| | | α | β | γ | δ | ε |
| | 1 | *b* | *d* | *a* | *a* | *d* |
| Environmental | 2 | *a* | *d* | *a* | *d* | *a* |
| States | 3 | *d* | *b* | *a* | *b* | *a* |
| | 4 | *d* | *a* | *b* | *d* | *a* |
| | 5 | *d* | *a* | *a* | *a* | ***b*** |

Response ε may have fallen into disuse because it does not provide favorable outcomes against any of the recently relevant environments 1, 2, 3, or 4. This would make it "ancient magic". Or, if it was never part of this firm's repertoire, behavior ε might have to be borrowed from the culture of another firm that has used it with success in a situation similar to "5". It might require an inter-firm alliance to acquire and transfer this "alien magic."

Johnson & Johnson's response to the Tylenol crisis was just such an ε behavior. Its unprecedented quick sharing of information with the public and recall of all its brands from store shelves not only assured the company's survival, but resulted in much-increased market share.

Conservatism, i.e., repeating one or more of the established responses α–δ, would reinforce the social cohesiveness of the organization. However, these simple tables show that in the face of the new condition 5, such conservatism would not allow the firm to survive.

# Information and mechanical models of intelligence

## What can we learn from cognitive science?*

Maria Eunice Quilici Gonzalez
Universidade Estadual Paulista (UNESP), Marília, Brazil

The impact of new advanced technology on issues that concern meaningful information and its relation to studies of intelligence constitutes the main topic of the present paper. The advantages, disadvantages and implications of the synthetic methodology developed by cognitive scientists, according to which mechanical models of the mind, such as computer simulations or self-organizing robots, may provide good explanatory tools to investigate cognition, are discussed. A difficulty with this methodology is pointed out, namely the use of meaningless information to explain intelligent behavior that incorporates meaningful information. In this context, it is inquired what are the contributions of cognitive science to contemporary studies of intelligent behavior and how technology may play a role in the analysis of the relationships established by organisms in their natural and social environments.

### 1. Introduction

Advanced technology has produced mechanical tools that allow cognitive scientists to simulate aspects of intelligent behavior with the use of computers (to create symbolic models) and robots. These constitute the basic tools of the cognitive synthetic methodology according to which mechanical models of the mind may provide good explanatory tools to investigate intelligent behavior. In this paper the mechanistic presuppositions of the synthetic methodology are analyzed by addressing the following questions: Is it appropriate to investigate intelligent behavior that involves meaningful information from a mechanistic point of view or does such an investigation require a systemic, non-mechanistic perspective? What are the advantages, disadvantages and implications of the synthetic methodology developed by cognitive scientists in the study of intelligence? These questions indicate the main line of the present investigation, which will be developed in three steps. The first (Section 2) discusses the cognitive science objective of building mechanical models of the mind based upon the synthetic methodology. Some limitations of this methodology will be indicated in the context of studies of autonomous intelligent action.

The second step (Section 3) considers the mechanistic approach to the mind, as developed by contemporary cognitive science, against the background of a philosophical scenario illustrated by the opposition Descartes / De La Mettrie. The aim is to show that the mechanistic supposition underlying the synthetic methodology, according to which intelligence can be understood in terms of mechanical laws, has been the object of intense debate amongst philosophers for centuries.

As stressed in Gonzalez and Haselager (2003), there is an historical conflict between the conceptions of philosophical (theoretical) and technical (practical) knowledge: Ever since the ancient Greeks there has been an over-valuation of intellectual abilities to the detriment of technical ones. The former were supposed to result from intelligent reflection developed by brilliant minds, whereas the latter were attributed to less intelligent creatures whose actions result mainly from mechanical bodily effort. As the mechanistic/non mechanistic debate acquired a new flavor in cognitive science with the expansion of information theory and the production of new technological artifacts, the concept of information is investigated, such as proposed by the Mathematical Theory of Communication (MTC) of Shannon and Weaver (1969) and developed by Dretske (1981,1991). This analysis is offered as a contribution to the debate about the advantages and implications of the contemporary mechanistic approach to the mind.

Finally (Section 4), I propose an alternative view to the synthetic methodology, stressing some characteristics of autonomous intelligent behavior from a systemic, non-mechanistic, perspective (von Bertalanffy 1969; Weinberg 1975).

## 2. The synthetic methodology of cognitive science: What is missing in the mechanistic approach to the mind?

The synthetic methodology constitutes one of the fundamental, distinctive strategies developed by cognitive scientists to investigate issues related to the nature of the mind and its implication for intelligent behavior. As mentioned, this methodology (which distinguishes Cognitive Science from other areas of investigation of the mind), presupposes that intelligent activity should be explained through the construction of mechanical models of the mind that could, in principle, simulate or reproduce its main features. These models are claimed to possess explanatory power in the sense that, by creating them, scientists supposedly demonstrate their understanding of the set of mechanical laws that could be responsible for the intelligent activity under investigation.

The synthetic methodology flourished about thirty years ago in the area of Artificial Intelligence (AI) with the help of functionalism and, more specifically, with the supposition that Turing Machines constitute appropriate instruments to build up symbolic (abstract) mechanical models of the mind that, in principle, could function independently of the environment and the material stuff of which it is built. As is well known in cognitive science, this symbolic approach to the mind — grounded as it is in Turing's (1950) hypothesis that intelligence can be understood in terms of the mechanical information-processing of symbols — was strongly criticized by philosophers such as Dreyfus, Searle, Baker, and many others.

One difficulty with the synthetic methodology, very much debated in the last century by philosophers, is that Turing Machines are developed in accordance with pre-established rules and their structures are pre-determined by the criteria of relevance which are supplied to them. This would not be, in itself, problematic if our object of investigation — intelligent behavior — did not require the understanding of the ways criteria of relevance are established in the first place (and given *a priori* to the machine) when we humans act in an autonomous and intelligent way. As Dascal (1990) pointed out, one of the main characteristics of intelligence is the ability to adapt pragmatically to changes in the context in which one is immersed. In this sense, he stresses that:

> ...The crucial question (for AI) is not that of 'representation of knowledge', nor that of 'providing (more) knowledge' to a system. Rather it is the question of designing systems that are not enslaved by something labelled 'knowledge', i.e., systems which are able to reject justifications that do not seem reasonable to them, and to select pragmatically even the criteria themselves of what is to be considered, in each context, as reasonable and 'relevant'... (Dascal 1990: 236).

Given that Turing Machines are subordinated to the criteria of relevance that are supplied to them it is difficult to see how they could be of great help in understanding and explaining the processes involved in autonomous intelligent action. It seems to be in this context that Dascal insisted that AI researchers should direct their attention to problems related to the possibility of developing systems that are not subordinated to pre-established rules:

> Researchers in AI should direct their attention to the question of whether it is possible to develop systems which are not subordinated to the knowledge and to the rules and criteria which are supplied to them *ex machina*, and if so, how. And they should not forget that this pragmatic aspect of knowledge derives from the public/social character of justification (Dascal 1990: 236).

The first part of Dascal's suggestion has been the subject of investigation by cognitive scientists, particularly those working with self-organizing neural nets and evolutionary robotics. For a long time they have been refining the synthetic methodology in order to preserve its mechanical presupposition about the lawful nature of intelligence without losing flexibility and autonomy. In the case of neural nets, as pointed out in Gonzalez (2000), self-organizing physical properties of memory have been stressed; and emergent properties of dynamical systems have been investigated in order to provide alternative explanations for the traditional functionalist AI approach to the mind. With the development of co-evolutionary robotics and Artificial Life (Nolfi and Floreano 2000; Murphy 2000), biological and collective properties of intelligent behavior are becoming more and more evident. In this scenario, the objective of building up mechanical models of the mind, based upon the synthetic methodology, has acquired "new" characteristics (similar to those present in the robotic models of the sixties). However, the second part of Dascal's suggestion — that cognitive scientists should not forget that "the pragmatic aspect of knowledge derives from the public/social character of justification" has been, to a great extent, neglected by mainstream cognitive science. The result is that a methodology that contemplates the analysis of the pragmatic systemic aspects of intelligence is still missing in the contemporary mechanistic approach to the mind.

In Section 3, it is argued that the limitations of the mechanistic approach to the mind have been the object of intense debate amongst philosophers for centuries. By inquiring about the limitations of the mechanistic approach it is intended to identify characteristics of intelligent behavior that seem to present meaningful pragmatic characteristics.

## 3. Meaningful information and intelligence

It was stressed in Section 2 that the assumption that intelligence can be understood and explained in terms of mechanical information processing is a fundamental axiom of the synthetic method of analysis developed by contemporary cognitive science. However, the mechanistic approach to the mind has also been the object of intense debate amongst philosophers for centuries. Thus Descartes, for example, in his correspondences of 1646 with the Marquis of Newcastle, makes explicit his opposition to philosophers who attribute intelligent abilities to animals. He admits that animals can, sometimes, do things better than humans, as in the case of animals that run or fly much faster than any human being. Nevertheless, he argues that this type of behavior only provides evidence that animals act in a mechanical, non-intelligent way "like clocks" that indicate time in a precise way by virtue of their physical structure and the disposition of their parts. He claimed that intelligent activity, in contrast to mechanical behavior, presupposes the ability to distinguish what is relevant from what is irrelevant, in different situations. Such a distinction is established in accordance with criteria given by our reason — the essential manifestation of our soul — and not by the dispositions of parts of our mechanical body or in accordance with a set of pre-established rules.

In opposition to Descartes, Julien Offray de La Mettrie argued, one century later, in his *L'homme machine*, that thought and intelligent behavior are just properties of organized matter, requiring only the right disposition of the parts of the body to exist. The more complex and structured the body-parts of an organism, the more developed its ability to act and think in intelligent ways.

In the 20th century, the opposition Descartes / De La Mettrie in relation to the mechanical nature of intelligence has also been addressed, among others, by Gilbert Ryle in *The Concept of Mind*. But this opposition acquired a different characteristic with the advent of new technological digital artifacts, such as sophisticated computers and robots, which allowed the synthetic methodology to be implemented in studies of cognition and action. In particular, De La Mettrie's hypothesis about the relationship between complexity and intelligence has been investigated from the informational perspective, according to which the complexity of a system can be measured in terms of the quantity of information necessary to predict its performance. The hypothesis brought to the fore in this perspective is that the more complex a system, the more information is necessary to predict its behavior.

The concept of information as applied in the synthetic methodology has never been clearly defined, but it is grounded on the notion of physical symbol processing (Newell and Simon 1972) and on the Mathematical Theory of Communication (MTC) (Shannon and Weaver 1969). In a nutshell, MTC provides a probabilistic characterization of information in terms of the reduction of uncertainty in the decision process involved in the choice of one message from a set of possible messages. In this context, uncertainty indicates a certain amount of randomness in addition to ignorance of the laws governing a sequence of events described by the messages in question. Information, in contrast, is related to what one could objectively predict, under specific conditions, once these laws are available and randomness is constrained.

As Shannon and Weaver stressed, MTC does not deal with the subjective elements of experience or with its possible meaning, but rather with the number of objective relations of dependency that may exist between events in a source. Even though these characteristics of MTC are, admittedly, inappropriate for studies of cognition and intelligence in general, this theory constituted a starting-point of a naturalist research program on the relationship between knowledge, information and action developed by Dretske (1981, 1991) and other philosophers concerned with the ontological dimension of information.

Inspired by Shannon and Weaver, Dretske defines information as "an objective commodity" or as "an objective, mind independent indicator of relations" (1981:55). However, in contrast to MTC, Dretske's naturalistic theory of information focuses on meaning and its relation to learning in the context of physical and biological structures. Thus, referring to the process of body conditioning, through which organisms (and even artificial neural networks) may start the acquisition of meaningful information, he claims that

> [b]y the timely reinforcement of a certain output — by rewarding this output, when, and generally only when, it occurs in certain conditions — internal indicators of this condition are recruited as causes of this output (Dretske 1991: 98).

Dretske argues that once these internal indicators are created, they may start to play an important role in the establishment of internal representations. These representations carry meaningful information that is relevant for the control of the organism's behavior:

> In the process of acquiring control over peripheral movements (in virtue of what they indicate), such structures acquire an indicator function, and, hence, the capacity for misrepresenting how things stand. ...This, then, is the origin of genuine meaning and, at the same time, an account of the respect in which this meaning is made relevant of behavior (Dretske 1991: 88).

According to Dretske's naturalist perspective, the world is full of information, but not full of meaning; only with the help of learning mechanisms (such as reinforcement and pattern association), motivations and reasoning related to internal representation and control of action, organisms (and maybe complex artificial neural nets) acquire the ability to process meaningful information. The question of whether the process of learning can be understood and explained by means of purely mechanistic laws is not very clear in this perspective. However, since he admits the possibility that a properly trained artificial neural network may be able to acquire internal indicators (through reinforcement and other forms of associative learning) that are relevant to its goal, then he would have to agree (as he seems to do) that this mechanical device could, in principle, process meaningful information.

The complicating factor in this view is that, despite considering the possibility that artificial neural nets may be able to process meaningful information (because they can learn and represent information necessary to reach a specific goal), he does not admit this possibility in the action of simple organisms, such as worms, caterpillars, moths, etc., which, according to him, adjust their behavior in accordance with simple mechanisms or tropisms:

> The moth's behavior is like so much of the behavior of simple organisms, tropistic. Tropisms are simple mechanical or chemical feedback processes or combinations of such processes that have the interesting property of looking like organized motivated behavior (Dretske 1991: 93).

The main reason why tropistic behavior does not count as intelligent autonomous behavior, which involves meaningful information, is because it is instinctive, automatic, genetically coded and, significantly, not modifiable by learning. Under these conditions, Dretske argues that very simple organisms do not deal with meaningful information, not because they are simple, but because their internal indicators (when eventually created) do not *represent* the world according to criteria of relevance established by reasons or intentions:

> Reasons are irrelevant to the explanation of this behavior, not because there is an underlying chemical and mechanical explanation for the movements in question (there is, presumably, some underlying chemical and mechanical explanation for the movements associated with all behavior), but because, although indicators are involved in the production of this movement, *what they indicate* — the fact that they indicate thus and so — is irrelevant to what movement they produce (Dretske 1991: 94).

In this context, the understanding of voluntary, autonomous action requires an apparatus of representational concepts that philosophers traditionally classify as belonging to the "intentional explanatory" domain. This involves purposes, beliefs, desires and all sorts of reasons that belong to the universe of internal, mental representation. As Ryle (1949: 73) stresses in his critique of what he calls the 'intellectualist' tradition, without this apparatus philosophers (following the Cartesian tradition) feel as if it "would be impossible to state what the qualifications are for membership of the realm of Spirit, the lack of which entails relegation to the realm of brute Nature".

Dretske's appeal to internal characteristics, in whose domain mental representations reside, in order to distinguish automatic, meaningless information from meaningful information brings us back to the opposition between Descartes and De La Mettrie, making us wonder what are the advantages of his informational approach for the understanding of autonomous intelligent action.

The above very summarized presentation of Dretske's ideas on the nature of meaningful information does not allow us to do justice to his attempt to explain the role of beliefs, desires, and reasoning in the intricate web of meaning that cognitively complex animals (such as humans) can create. However, for the present purpose it could be argued that one advantage of his naturalist informational view (as compared to the traditional Cartesian position, for example) is the emphasis given to the body and the environment in the production of meaningful information. As Dretske points out, his theory is close to Block's theory of meaning, according to which

> the meaning of internal elements is a combination of (1) their relations (usually causal or informational relations) to the external situations they represent and (2) their functional (or conceptual) role in the production of output (including their internal relations to each other) (Dretske 1991: 150–151).

Dretske also stresses the holistic and dynamic character of meaning, which may be initiated with a relatively simple memory structure — such as that of a neural network that learns how to identify patterns through the composition of indicator relations. As the passage above indicates, the meaning of a network of associated meaningful relations may give place to a variety of other indicator relations generating new dimensions of meaning.

One important consequence of this informational perspective for the understanding of intelligent behavior involving meaningful information is that the distinction between humans, other animals and machines is essentially a matter of degree. The question of what makes a physical system an autonomous intelligent agent should probably be answered in terms of the system's ability to develop high levels of meaningful information whenever new and more complex situations arise in the system's environment, requiring new pragmatic solutions for its maintenance. But, how would this happen? What could possibly be the production mechanism of new, relevant (meaningful) information that would help with the maintenance of the system? Dretske agrees that, at the biological level, it is a complete mystery how new high levels of meaningful information are produced out of structures already available:

> … we could imagine a high-level switching mechanism, sensitive to other channels of information and to collateral motivation variables, controlling the activation of already established (through learning) dispositions. But the production of novel responses, the essence of intelligent behavior, cannot be thought about in such simple, mechanical terms (Dretske 1991: 141).

Admittedly, the history of mechanical reinforcement, which gives place to conditioned activities related to control mechanisms, seems to be suitable for the analysis of simple behavior of organisms and artificial creatures, such as neural networks and reactive robots. Yet, it does not seem appropriate for accounting for the main characteristics of intelligent autonomous action.

Consider, for example, the case of a reactive robot-environment system, which incorporates circular feedback and self-organizing mechanisms (as in Verschure 1996). As described in Gonzalez and Haselager (2004), this system contains a light-sensitive robot, known as Roboser, which moves around, propelled by external sources of light, on a flat surface. As it is designed, Roboser has a mechanism of aversion to any sort of physical contact, so that, whereas it is attracted to light, it will run away whenever its body touches any object. When incorporating the goal to reach a specific point, for example, it will move around in order to find its own way, creating relatively stable habits.

Now, would it be correct to suppose that Roboser is an autonomous intelligent system? The immediate answer is no, because it is a pre-wired system with no freedom to choose any particular item available in the world. But, what if Roboser could learn and acquire a background of interconnected relations in the way indicated by Dretske? This still would not do, because Roboser would not have a criterion of relevance for distinguishing between appropriate and inappropriate habits. This leaves us with our initial question of the advantages and implications of the mechanistic (representational and informational) presuppositions of the synthetic methodology for the understanding of this matter. In the next section, I will sketch a systemic approach to information that contemplates a non-mechanistic view of intelligent behavior.

## 4. Information flow and the implicate order: An alternative to mechanicism

So far we have examined two basic presuppositions of the synthetic methodology of cognitive science: (1) mechanical models may provide good explanatory tools to investigate the mind and (2) intelligence can be properly understood in terms of objective information processes that involve learning mechanisms directed to the establishment of representational control mechanisms. As an example of (2), we summarized Dretske's model of meaningful information and indicated its apparent limitations in the investigation of intelligent autonomous action.

We believe that one problem with Dretske's view is that, despite stressing the holistic character of meaning, his model reflects a fragmentary way of looking at the nature of information, mind and reality, grounded on the representationalist tradition. The MTC roots of his model, based upon the ideal of information understood as an objective commodity — as "a mind independent indicator of relations" — indicates the initial trace of this fragmentation. Why should the mind be excluded from the informational universe? Would this ideal not defeat his claim about the holistic character of meaning that involves a dynamic, complex network of relations? What would be the main difficulty with a truly holistic view of information that includes the mind in the very origins of the process of forming/grasping indicator relations constitutive of information?

A short answer could be that, since meaningful information involves, according to the representationalist view, the existence of mind-like systems for the production of (learning) representational mechanisms, the assumption that these systems exist would introduce a circularity that would destroy the whole project of explaining, in an objective way, the emergence of meaning in the first place. Nevertheless, if one does not subscribe to representationalism, this problem does not necessarily arise.

In what follows I outline an alternative view of representationalism and of mechanicism, according to which the world is full of meaningful information that does not require representations to be grasped, but rather the organism's perception of an 'implicate order' (Bohm 1980), to which it belongs. This view is anchored in Gibson's concept of *ecological information* understood as a dynamic and intricate web of 'affordances' — a resource of the world that has the potential to enable organisms to encounter opportunities for action (Gibson 1979).

Gibson's notion of *ecological information* can only be understood in a context of agent-environment couplings, in which environment and agents co-evolve as a unity that cannot be fragmented. In this holistic process of evolution, the dynamic interaction between body and environment *mutually* constrains the perception-action of organisms, giving place to relatively invariant (physical, biological and social structures) meaningful resources that have the potential to facilitate, as in a chain, the use of other resources. As environment and agents evolve as a unity, this rich variety of 'invariants' constitutes an ordered flux expressing a myriad of patterns that are meaningful and ready to be grasped, without representational mediations, by attentive organisms in their specific process of perception-action.

From a similar perspective, Stonier (1997: 120) argues that meaningful information exists in a spectrum that varies from very simple to complex levels: "At the high level, meaning leads to understanding and further analysis, while at the low end, meaning involves something as simple as the ability to pick out an item from among an array of items". As an illustration of this view, he considers an imaginary situation in which a whole team of workers, with different habits and cultural backgrounds, is engaged in a house cleaning service. While doing their jobs, one of them discovers a piece of a jigsaw. Assuming that jigsaw puzzles are not familiar to this cleaner, that piece would not mean too much for him. However, when taken to someone whose cultural background includes jigsaw puzzles, this piece is immediately recognized as a piece of a jigsaw and is taken to the mistress of the house. As she has the information that her son has being working with a puzzle for a while, she takes the lost piece upstairs, where he has practically finished putting the puzzle together.The boy immediately recognizes the piece, as soon as his mother shows it, as 'a part of the sky' that was missing in his puzzle.

The moral of the above thought experiment is that "...the same piece of information may increase in meaningfulness depending on who looks at it" (Stonier 1997: 120). Stonier stresses that this meaningful aspect of information does not depend specifically on the human mind, in the sense that it is not an exclusive property of humans. It is rather

> interrelated phenomena which extend through all of nature — patterns which range from subatomic structure to the human mind and human society ...this pattern consists of the manifestation of order, to a greater or lesser degree, in everything we perceive, create, or conceive (Stonier 1997:1).

Investigations on the nature of 'order' and its myriad of manifestations in the experience of organisms are at the root of the ecological approach to information; it is present, for example, in Gibson (1979), in Bohm (1980) and in Bateson's (1979, 2000) famous characterization of information as "the difference which makes a difference" (2000: 315). The emphasis on the relation between organisms and environment is also present in Wiener's definition of information as "a name for the content of what is exchanged with the outer world as we adjust to it, and make our adjustment felt upon it" (Wiener 1948:19).

A contemporary development of Wiener's ideas has produced advances in the dynamical, non-representationalist branch of cognitive science, especially in the robotics reactive paradigm of Brooks (1999), Murphy (2000), and others. As stressed in Gonzalez and Haselager (2005), this paradigm occupies a special place in the spectrum of AI research. Instead of being immobile, knowledge driven, inference machines, reactive robots operate without any explicit central knowledge consultation, inferential processes or model construction. Instead, they consist of behavioural layers, which connect input directly with output, constituting a behavioural pattern or habit. The layers do not consult each other by means of representation exchange, but rather compete for dominance by means of inhibition or suppression mechanisms. The overall behaviour exhibited by the robots is an emergent result of the interaction between the layers, both among themselves and with the environment. In this scenario, robots are capable of interacting directly with the world, without any representational mediators, solving problems by means of their hardwired behavioural capacities or habits.

Even though the emphasis given by the reactive robotics paradigm to the body and environment brings its researcher much closer to the ecological approach to the mind than the traditional representationalist AI, the results so far obtained in studies of intelligent actions seem to be very poor. This is because reactive robots do not have criteria of relevance of their own to move around and to select their own path in the environment. Also, as the robot-environment couplings are, in general, pre-wired, they cannot modify and be modified by their surroundings. One consequence of this is that they do not, for example, "step out" of disturbing, unpredictable situations in order to observe that something unusual may be blocking their habitual way of dealing with the environment (Gonzalez and Haselager 2005). Great attention has been given to this kind of problem in evolutionary robotics, as developed, for example, by Nolfi and Floreano (2000) and others that make use of genetic algorithms in biology and self-organizing mechanisms, in order to investigate more flexible models of intelligent autonomous behavior.

Research on evolutionary and co-evolutionary robotics is just starting, but it is possible to expect some results in the coming decades in terms of the creation of sophisticated mechanical models of the mind that incorporate more and more new advanced technology. This possibility mandates inquiring into the presuppositions that are at the basis of these models and their implications for the understanding of meaningful information and its relation to intelligence. The great fear (shared by many humanists) is that the development of mechanical models may blur the distinction between mechanical and systemic order, implicit in the traditional studies of intelligence and autonomous action.

One characterization of these two basic notions of order — mechanical and systemic — is, according to Bohm (1980: 173), the following:

> The principal feature of this [mechanistic] order is that the world is regarded as constituted of entities which are outside of each other, in the sense that they exist independently in different regions of space (and time) and interact through forces that do not bring about any changes in their essential natures. The machine gives a typical illustration of such a system of order. Each part is formed (e.g., by stamping or casting) independently of the others, and interacts with the other parts only through some kind of external contact.

In contrast to mechanistic order, he stresses the systemic character of living organisms, i.e., the fact that "each part grows in the context of the whole, so that it does not exist independently, nor 'interacts' with the others without itself being essentially affected in its relationship" Bohm (1980: 173).

Researchers interested in the explanation of this systemic notion of order, such as Bateson (1979, 2000), Bertalanffy (1969), Debrun (1996), Haken (1999, 2000), Weinberg, (1975), and others suggested a method of investigation of the dynamics of complex systems grounded on the Theory of Self-Organization (TSO). This focuses on studies of processes through which new forms of order emerge mainly from the dynamic, spontaneous *interactions* between elements without the action of an external central controller.

The systemic methodology started with studies of the organization of complex structures, such as those related to living organisms and society, and expanded to physical, chemical and informational structures. In this context, information can be defined as a self-organizing process that allows the expansion of patterns of action for organisms situated in their specific environment. This process acquires meaning as a consequence of the systemic relations established by organisms in their natural environment. In line with Stonier's hypothesis mentioned above, it is suggested that meaningful information plays a role in action, a role that varies from simple to high levels, depending on the degree of complexity of the patterns in question. In the case of human beings, these patterns involve, among others, biological, pragmatic and socio-cultural dimensions.

The above definition of information is proposed as an alternative to the mechanistic representational characterization of information. The systemic notion of order implicit in it was inspired in Bohm's concept of *implicate order* (from the Latin root meaning "to enfold" or "to fold inward"):

> In terms of the implicate order one may say that everything is enfolded into everything. This contrasts with the *explicate order* now dominant in physics in which things are unfolded in the sense that each thing lies only in its own particular region of space (and time) and outside the regions belonging to other things (Bohm 1980: 177).

But if everything is enfolded into everything, how could we distinguish relevant from irrelevant information? As Bohm indicates, the word 'relevant' derives from the verb 'to relevate', whose meaning is : "to lift into attention, so that the content lifted stands out 'in relief': When a content lifted into attention is coherent or fitting with the context of interest this content can be said to be relevant" (1980: 33). In contrast, he stresses, when a content lifted into attention is incoherent, or does not fit in a context, it is said to be irrelevant.

The sentence "This watch does not run, even though I used the best butter" (from Lewis Carroll's dialogue between the Mad Hatter and the March Hare) is suggested by Bohm as an illustration of the above characterization: "It lifts into attention the irrelevant notion that the grade of the butter has bearing on the running of watches ..." (ibid.). But, again, how do we know that the grade of the butter is irrelevant to the running of watches? Could there be an algorithm for a mechanical apprehension of these data? Bohm's answer is that the irrelevance in question becomes evident from the fact that the idea of using the best butter does not fit the context of the actual structure of watches. This inference requires a context in which agents situated in a socio-cultural environment have basic skills concerning the running of watches.

Questions related to the possibility of creating an algorithm for the mechanical apprehension of relevant information bring us back to the mentioned opposition Descartes/De La Mettrie. Our guess is that organisms, in order to survive, developed basic systemic skills for the body's self-organizing process of adjustment to the environment. As organisms and environment co-evolved these skills were further developed giving place to the emergence of criteria of relevance that allowed more refined adjustments according to social demands. In this sense, as suggested by Bohm, the notion of relevance presupposes "an act of perception of very high order". He claims that this perception is not governed by strict rules:

> Clearly, the act of apprehending relevance or irrelevance cannot be reduced to a technique or a method, determined by a set of rules. Rather, this is an art, both in the sense of requiring creative perception and in the sense that this perception has to be developed further in a kind of skill (as in the work of the artisan) (Bohm 1980: 33–34).

Bohm's characterization of the difficult notion of relevance in terms of artistic skills suggests that pragmatic and social aspects are at the center of our notion of intelligence; the nuances of these skills cannot be reduced to a technique or explained by means of a synthetic methodology because they require the proper use of meaningful information However, there is a fear, particularly among humanists, that the growing fascination exerted by new advanced technology and the spreading of the synthetic methodology to issues that concern meaningful information may lead to neglecting the importance of art and its biological, social and historical roots, diminishing the meaningfulness of abilities that lie at the center of our culture and existence. Real as this fear may be, it does not take into consideration that we might learn from cognitive science that certain kinds of intelligent behavior, such as chess and flying performances, for example, can be explained in terms of mechanical strategies or embodied successful habits.

A way out of this dilemma concerning the possible impact of new advanced technology on issues that concern meaningful information and its relation to studies of intelligence seems to be the broadening up of perspectives: Instead of confronting two different views, it could be helpful to adopt Bohm's thesis that

> man is continually developing new forms of insight, which are clear up to a point and then tend to become unclear. In this activity, there is evidently no reason to suppose that there is or will be a final form of insight (corresponding to absolute truth) (Bohm 1980: 177).

A perspectivist, non-reductionist approach to intelligence seems to be the best way to deal with such a complex subject.

## 5. Concluding remarks

In his very creative paper "An imitation of life", Grey Walter mentions legends, common in the "dark ages", of living statues and magic pictures used as models of an enemy. These statues were supposed to embody, somehow, the enemy's soul in a way such that injury to the model would be reflected in the original. He argues (Walter 1950: 42) that

> [i]dolatry, witchcraft and other superstitions are so deep-rooted and widespread that it is possible that even the most detached scientific activity may be psychologically equivalent to them; such activity may help to satisfy the desire for power ...or to compensate for the flatness of everyday existence.

In this context, Walter comments on the intense modern interest in machines that imitate life and proposes a model of mechanical tortoises that possess very simple electronic nerve cells and somehow exhibit what he calls "free will". His purpose with these models is to try to shed some light on the studies of processes concerning the structure of life. By helping, with the future efforts of cognitive science, to take away man from the center of creation, he introduces a very simple (and why not to say — naïve) strategy to investigate "free will" from which one may conclude that, after all, the fear of mechanicism is perhaps not fully justified. Indeed, we might learn from his little robots that certain kinds of intelligent behavior can be explained in terms of embodied successful habits. "Free will" may be, as Walter claims, an emergent property of informed actions of embedded agents which care for their survival.

Considering the possibility that the process of the body's adjustments to the dynamics of life provides criteria of relevance for the orientation of basic actions responsible for survival, why is it that cognitive science has not investigated further this process in the study of autonomous intelligent action? The provisional conclusion is that the main difficulty with the above suggestion is that it would be beyond the reach of a purely mechanistic approach to the mind, as developed by robotics and other branches of cognitive science. Systemic, not purely mechanical, aspects of biological and socio-cultural flows of information seem to provide additional constraints that need to be taken into account for a proper analysis of meaningful information related to real living organisms in the domain of intelligent autonomous action. A new agenda for cognitive science could then be suggested, based on Bohm's theory of 'implicate order'. This should include a holistic approach to the mind according to which the mechanical modeling strategy is just one part of a complex set of tools for the investigation of intelligence and action. A further suggestion is that, once (and if) the fascination with and fear of mechanicism is overcome, cognitive scientists could allow themselves to consider the complex dynamics of body-environment couplings and the impact of new advanced technology on issues that concern meaningful information.

## Note

* I would like to thank Mariana Claudia Broens, Ramon Capelle, Pim Haselager, José Arthur Quilici Gonzalez, Thiago Carreira Alves Nascimento, and my research fellows from CLE and GAEC for their suggestions and support during the preparation of this paper. I would also like to thank Andrew Allen for the English corrections, and FAPESP, FUNDUNESP and CAPES for funding this research.

## References

Bateson, G. 1979. *Mind and Nature: A Necessary Unity*. New York: Cambridge University Press.
Bateson, G. 2000. *Steps to an Ecology of Mind*. Chicago: The University of Chicago Press.
Bohm, D. 1980. *Wholeness and the Implicate Order*. London: Routledge and Kegan Paul.
Bertalanffy, L. von. 1969. *General System Theory*. New York: Braziller.
Brooks, R.A. 1999. *Cambrian Intelligence: The Early History of the New AI*. Cambridge, MA: The MIT Press.
Dascal, M. 1990. "Artificial intelligence as epistemology?" In E. Villanueva (ed), *Information, Semantics and Epistemology*. Oxford: Blackwell, 224–241.
Debrun, M.A. 1996. "A idéia de auto-organização (The idea of self-organization)". In M. Debrun, M.E.Q. Gonzalez, and O. Pessoa Jr. (eds), *Auto-organização: Estudos Interdisciplinares (Self-organization: Interdisciplinary Studies), Vol. 18*. Campinas (Brazil): Centro de Lógica e Epistemologia, UNICAMP, 1–23.
Dretske, F. 1981. *Knowledge and the Flow of Information*. Oxford: Blackwell.
Dretske, F. 1991. *Explaining Behavior*, 3rd ed. Cambridge, MA: The MIT Press.
Gibson, J. 1979. *The Ecological Approach to Visual Perception*. Boston: Houghton-Mifflin.
Gonzalez, M.E.Q. 2000. "The self-organizing process of distributed information: A way out of the mind-body problem?". In *Proceedings of the Fifth Brazilian-International Conference on Neural Networks*. Rio de Janeiro, 153–166.
Gonzalez, M.E.Q. and Haselager, W.F.G. 2003. "Multicultural co-existence and e-learning: Any place for creativity and self-organization?". In T. Kobayashi (ed), *E-learning beyond Cultural and Linguistic Barriers: Co-existence and Collaboration. Research Report of the National Institute of Multimedia Education*. Chiba, 53–67.
Gonzalez, M.E.Q. and Haselager, W.F.G. 2005. "Creativity: Surprise and abductive reasoning". *Semiotica* 153: 325–341.

Haken, H. 1999. "Synergetics and some applications to psychology". In W. Tschacher and P.P. Daualder (eds), *Dynamics, Synergetics, Autonomous Agents*. London: World Scientific, 3–12.

Haken, H. 2000. *Information and Self-organization*. Berlin: Springer.

Murphy, R.R. 2000. *Introduction to AI Robotics*. Cambridge, MA: The MIT Press.

Newell, A. and Simon, H. 1972. *Human Problem Solving*. Englewood, NJ: Prentice Hall.

Nolfi, S. and Floreano, D. 2000. *Evolutionary Robotics: The Biology, Intelligence and Technology of Self-organizing Machines*. Cambridge, MA: The MIT Press.

Ryle, G. 1990. *The Concept of Mind*, . 3rd ed. London: Penguin Books.

Shannon, C. and Weaver, W. 1969. *The Mathematical Theory of Communication*. Chicago: University of Illinois Press.

Stonier, T. 1997. *Information and Meaning*. London: Springer.

Turing, A. 1950. "Computing machinery and intelligence". *Mind* 59: 236–245.

Verschure, P.F.M.J. 1996. "Minds, brains, and robots: Explorations in distributed adaptive control". In *Proceedings of the Second Brazilian-International Conference on Cognitive Science* UENF, Campos, Brazil, 97–106.

Walter, G.W. 1950. "An imitation of life". *Scientific American* 182(5): 42–46.

Weinberg, G. 1975. *An Introduction to General System Thinking*. New York: John Wiley and Sons.

Wiener, N. 1948. *Cybernetics, or Control and Communication in the Animal and the Machine*. Cambridge, MA: The MIT Press.

# Is cognition plus technology an unbounded system?

## Technology, representation and culture

Niall J.L. Griffith
University of Limerick, Ireland

The relationship between cognition and culture is discussed in terms of technology and representation. The computational metaphor is discussed in relation to its providing an account of cognitive and technical development: the role of representation and self-modification through environmental manipulation and the development of open learning from stigmery. A rationalisation for the transformational effects of information and representation is sought in the physical and biological theories of Autokatakinetics and Autopoiesis. The conclusion drawn is that culture, rather than being an intrinsic property of our human phenotype was learned and that cultural cognition is an information transforming system that is inadequately characterised by notions of parameterised deep-structure and that it is an open and potentially unbounded informational system.

### 1. Introduction

It has been suggested that the interaction of cognition with the environment is so extensively mediated by technology, that changes and developments in technology produce effects that change the nature of cognition itself.

The mutuality of technology and human activity is so mundanely present in human experience as to be unremarkable. However, technology also permeates more abstract concepts and ideas that frame our view of the world and human nature, reinforced by the imaginative projection of our nature onto technology and its use in art (Pacey 1999).

One example of the close relationship between technology and our view of human nature is the mechanistic account of 'organisms' that developed from the Renaissance through the Enlightenment (e.g., mechanical automata), into the abstract information theoretic account of cognition in the computer metaphor of mind.

Apart from its mechanisation, this view of cognition focuses exclusively on the phenotypic 'hardware' capacity or function of the brain, latterly situated or grounded in the body. However, a complementary view of cognition is that it includes the systematic organisation of ideas, tools and beliefs encapsulated and conserved by culture. This perspective suggests that a significant aspect of the interaction of mind and reality lies in the transactions between the ideas that an individual (brain) receives culturally as well as naturally and its ability to transform and reformulate these received ideas. This emphasises the brain's flexibility as a receiver, carrier and transformer of ideas as well as its capacity to respond to and generate ideas directly from its experience of the natural world. It implies that cognition does not distinguish natural and cultural identities and processes.

Human cognition is understood through our participation in an integral system that involves the interaction of three elements: an intrinsic mechanism that operates locally and is actuated in the brain; a global system, culture; and the natural world that the individual and culture operate within and with which they are in continual dialogue.

The individual, somatic, genetic and evolutionary aspect comprises the innate capacities and capabilities of our species' cognitive hardware; while the collective, social, cultural, abstract, historical aspect, is focused on human learning and development. The 'hardware' phenotypic view is conservative in the sense that it is concerned with the specification of cognition in the capacities of an anatomical adaptation, which is co-terminal with our species' phenotype. The cultural perspective is concerned more with the properties of the system overall and the capacity of its component 'mechanisms' to transform the system they instantiate into new modes, i.e., the systemic informational aspects of social cognition. While this division tends to polarise views into oppositions such as nature and nurture, the alignment of 'closed' with 'innate' and 'open' with 'culture' is not straightforward. Evolutionary views recognise the adaptability of individual traits while some social theories emphasise interpretations of human social development in which technology in particular plays a strongly determining role (Heilbroner 2001). The divide between views emphasising the importance of individual hardware and collective culture is considerable. Accounts of cognition that acknowledge the importance of culture are essentially dualistic; recognising that cognition involves endogenous and exogenous aspects; conservation and transformation.

A basic view of technology is that it reflects interactions between phenotype and the world in an objectified form. There are physical tools, and tools for thought. Given that people invent and use tools and that tools themselves carry within their form indicators of their use and meaning (Norman 1987), it seems natural that cognition itself is in some way changed by the significations contained in artefacts as well as by the behaviour they facilitate and define.

Knowledge and understanding make us 'fitter', as a species; the information we use and the instruments that derive this information are tools that 'fit in' with the world. Technology informs our view of the world, and ourselves; but does this change cognition *per se*, or just 'the way that we think'; i.e., what we think about? The emergence of writing and the world-wide-web are examples of technology playing midwife in reformulating cognitive modes and the emergence of incremental orders of organisational complexity, i.e., of a new 'kind' (Goody 1986; Rheingold 2000). The paradigm shifts in science (Kuhn 1962) may accompany such revolutions. But do they constitute a change in the nature of cognition?

It is clear that any discussion of these issues is likely to involve strong assertions about the 'nature' of cognition. To reiterate these two views: one is that cognition is an evolved operational capacity specified and confined within the species-specific hardware functionality of the brain, which it cannot transcend. In this view we are the above averagely clever great ape. The other view is that the endogenous (endosomatic) hardware specification of the human brain is involved in interaction with the information it derives from the world mediated via exogenous (exosomatic) cultural organisation and conservation. In this view, while operational in its origin, our intelligence is more than grounded in our experience of the world. It is embedded in the same physical laws and reflects or expresses the same underlying principles. By implication, while we may never fully understand ourselves we cannot rationally put a limit on what human culture might achieve in terms of it's collective understanding of the universe and our place in it.

This view seems to be an interesting perspective, and the aim here is to discuss some of the evidence for it. The history of technology is indicative that this dynamic interaction is the core devolvement of technology on cognition and that because of it the nature of cognition as a system changes. This does not, however, imply that human nature changes. Any capacity of cognition to shift its *modus operandi* in radical ways is taken to be intrinsic to our nature. Secondly, the aim is to consider the question of whether cognition as a systemic whole, involving the brain, culture and the world is an unbounded system.

It is axiomatic that the dynamic of human technological development lies in the ongoing interaction between use, function and activity. The design and use cycle is experimental, and iterative. Tools change and transform their function in the activities that use them.

The impetus to create tools is an intention prior to as well as coextensive with the behaviour they mediate. There is a tension between the significance we ascribe to the processes of internalisation and externalisation (learning and creation) as parts of the design-use cycle. We see ourselves either as complex reactive, i.e., behaviouristic entities caught in the mechanistic, unconscious operation of stimulus and response, or as conscious, intentional actors who can wilfully dissociate the internal focus of our imagination from actual action (Dennett 1991; Searle 1980). We think of ourselves as creative rather than reactive; yet paradoxically we often replace reaction with passion, i.e., unrealised action.

The separation of thought from action is associated with planned actions and making tools. Tools reflect our being decoupled from the world as much as our contiguity with it. Being able to conceive of actions independently of their actuation may result in a breakdown between intention and action (Ramachandran 2003), but this invalidity can only occur if intention does not automatically lead to action. The separation of imagination from action may facilitate the relationship between 'passing' internal and 'constant' external representation that engenders the more general duality of cognition and culture including our ability to assess evidence before acting.

Our capacity to defer action is closely associated with our view of ourselves, and the operational success of technology and science. How we understand these historical developments to have occurred should be consistent with our understanding of mind. The dominant contemporary model of mind is computational. Can a computational model of mind offer a persuasive account of human technological development?

## 2. The computer metaphor, specification and learning

The computational metaphor of mind is appealing because we anthropomorphise machines. When the machine involved is itself a programmable information transducer that can simulate attributes of cognition — induction and rule based behaviour — the identification is even stronger. The computational metaphor reflects the power of a technology to define our view of ourselves. The question is whether it is a good model.

If the brain is a mechanism that instantiates a Turing Machine (Newell and Simon 1976; Turing 1950), then, insofar as we may conceive of different Turing Machines, or programs of that class defined by the notion of a Turing Machine, transforming from one Turing Machine into another or reprogramming cannot be construed as a change in nature. The computational metaphor of mind is a static view of human cognition (Searle 1980). Computers cannot re-specify their own code. As alchemy failed to transform base metal into gold, so also the invocation of a programme has failed to instantiate mind.

The most compelling reason why machines cannot reformulate themselves is that they are not capable of conscious intention. Searle's (1980) Chinese Room analogy argues that conscious internal/meaningful states constitute intelligence and that meaning cannot arise simply from the form and syntax of symbols in relation to the world. Intention seems to implicate consciousness as a significant aspect of the brain's causal properties. However, Searle's treatment of the meaningfulness of internal states brings into focus how meaning arises, and in particular the self-referential nature of the learning through which new meaning arises. One crucial question for the computational metaphor is in what ways, if any, a Turing Machine can learn.

In the Chinese Merry Go Round argument Harnad (1990) considers whether a symbol system can learn to ascribe meaning to symbols — i.e., can syntax (by itself) give rise to semantics. Harnad asks us to imagine learning Chinese as a first language from a Chinese/Chinese dictionary and argues that this is impossible because the meanings within the system are extrinsic rather than intrinsic to it.

The Turing Machine as a model of in-the-head-cognition seems to be weak (Searle 1980; Putnam 1991; Fodor 2001). However, we can ask whether Turing Machines interacting with a culture and environment are capable of fundamental change. What if meaning arises through informational bootstrapping in the exogenous superstructure of culture, i.e., from the interactions between machines and environment?

While culture impresses itself onto the individual through acculturation it does not follow that meaning itself originated in this way. Rather, it suggests that culture involves organisms that already possess symbols whose meaning is intrinsic to them, as it is only thus that they can share meaning with other individuals. However, it is possible to conceive of 'intermediate' pre-cognitive automata that reflect the emergence of meaning and representation from syntactically driven exchanges. Arguably, stigmery[1] (Theraulaz and Bonabeau 1999) exemplifies a process that expanded self-awareness via Craikian automata (Johnson-Laird 1983).[2]

This suggests that rather than the brittleness of computation, the impetus underlying the evolution of mind selected for generality and openness as well as the specific phenotypic adaptations proposed in evolutionary accounts (Barkow et al 1992; Mithen 1996). Viewed from this perspective, human intelligence is an evolutionary attempt at an open-ended capacity for behavioural self-specification, namely the capacity to learn. This openness implies a decoupling of mechanism and action and this seems to be manifest in the human ability to represent thoughts and actions, i.e., to internalise and externalise cognitive as well as real objects and events, and the coextensive ability to dislocate imagination from action (Ramachandran 2003), allowing intentionally derived ideas and concepts to augment and supplant conditioned reaction to stimuli.

Considering the limitations of the Turing Machine model highlights that human cognition learns about causal structure from the world and consequently is itself redefined by the 'information' it extracts and organises. If we are to be persuaded that human cognition can change fundamentally we must convince ourselves that it is not bound by the physical mechanism that facilitates learning, i.e., that learning is behaviourally open.

## 3. Technology, culture and learning

What is the relationship between technology, culture and learning? Learning in the most general sense is the capacity to change behaviour in the light of experience. Evolution itself is one form of learning. For many species learning is restricted to development in which an organism fine-tunes its maturing capacities. However, more radical adjustment of specification to circumstances is also manifest not only in acculturation, which is fairly high-level, but also in the space-time dependencies of topological order (Edelman 1988), as well as informational reprogramming during development. For example, if the developing visual pathway of a ferret is redirected to the auditory cortex, it develops into a functionally viable visual cortex, albeit morphologically imperfect (von Melchner et al 2000). Thus, an area of a ferret's cortex, which might be supposed to be genetically determined, is plastic enough for its specification to be overridden by 'experience'. Organic hardware can ontologically establish its function.

Prior to the emergence of our own species, 'animal' culture can be viewed as a complex form of stigmery (Theraulaz and Bonabeau 1999). The classic exemplar of stigmery is the ant colony, where diverse individual activity results in environmental modification that prompts other, apparently disconnected individual behaviour that coalesces in the coherent collective action that characterises 'the anthill'.

Like stigmery, animal culture is essentially static and consistently defines rather than innovates, engendering organisation via behaviour. It is closed in as much as, for example, the ant's specification determines what emerges from a 'stigmergetic' exchange. Stigmery does not involve learning, except as the emergent organisation was learned through the evolution of the stigmergetic response itself.

Animal cognition and learning seem to be easily aligned with speciation. However, once ancestral hominids appear on the scene it becomes more problematic to explain change in cognition with reference to speciation alone. From the emergence of Australopithecines the issues shift from what cognitive properties are intrinsic to how far the brain has advanced in supporting cultural plasticity and innovation (Mithen 1996; Klein 1999). With the appearance of the hominids, culture and particularly technology, becomes prominent, consistent, and subject to change. The increasing complexity of technology suggests that our capacity to learn in this way may have been selected for. One question is whether the relationship between learning and culture can itself evolve.

As a species we are learners and information conservers. While we are not unique in learning, it is a highly developed aspect of our cognition. Moreover we are general learners. By this we mean that our learning is relatively unconstrained as to its means and ends, which are culturally negotiated priorities and possibilities, e.g., astronomy in the Renaissance and genetic engineering currently. Culture both instantiates or actuates what is learned as well as preserving and guiding it. Culture and technology conserve, and consequently cognition as a systemic property exists *de facto* in the relationship between our behaviour and external memory, as well as in the activity and (internal) representation within an individual. The symbiosis of action and object in internalisation and externalisation facilitates both the abstraction of the world and the objectification of abstractions in behavioural practices, rules and tools (Vygotsky 1986; Leroi-Gourhan 1993; Ingold 1993). While endosomatic memory is evanescent, external memory is objective — it is physically constant in things and psychologically constant in the continuity of learned, shared modes of cultural behaviour; rules, practice, laws, etc. Cultural mores and technologies present themselves to their users as natural, objective and truthful. For example, capitalism is currently rationalised through the belief that market forces are 'natural'. There are good reasons for thinking that (in evolutionary terms) cultural practices and beliefs would be less effective if they were not prescriptive. Yet a culture is a system of beliefs and actions that vary in utility, truth and objectivity.

General learning and culture seem to be closely connected. Culture can be seen as a natural development from more focused learning within specific domains (Mithen 1996) and may mediate general learning. Being an isolated general learner involves a conundrum. How does such a mechanism learn generally what is worth learning? Being able to make this judgement is advantageous given that any learning mechanism is imperfect and that any significant learning — invention or creation — involves risk. Also, unconstrained general learning is not optimal, because it re-invents the wheel.

For an isolated learner (mechanism or organism), this is a problem without a solution. Trying to learn what is worth learning leads to an infinite regress of trying to learn how to judge what is worth learning by trying to learn the best way of judging a mechanism that judges what is worth learning and on and on *ad infinitum*. There is no *deus ex machina* to determine success.

The relationship between learning and re-invention is however close. Pupils are given solved, known problems to mimic or rehearse — i.e., being trained *how to learn* does involve re-inventing the wheel. Culture and learning are bound together in mimicry and innovation. Crucially, in human culture conservation and innovation are bound to externalisation.

If solutions are remembered externally and culturally then three things can follow: a) good solutions are preserved and can be copied, b) learners don't waste energy re-inventing prior solutions, c) as a consequence of a) and b) learning is a structured process that is guided by its cultural context and conditions; practices, rules, organisation, and behaviour. In all societies culture controls and directs skills through education. This may favour innovation or conservation. Contemporary science celebrates novelty, but there are fewer brownie points for confirming a theory and fewer again for rediscovering it. The Bindibu people of Western Australian restrict to initiates the preparation of spear shafts from rare, fragile roots (Thomson 1964). Industrial science restricts use of (expensive, delicate) equipment to trained technicians and scientists. The allocation of the superabundance of resources available to science is founded on a parsimonious principle that reflects its evolutionary inception.

It is difficult to separate in any meaningful way culture from learning. The externalisation of new solutions and the internalisation of old is dynamic and involves conservation that encapsulates the potential for innovation. Technology has benefited from this dynamic as well as being a vehicle for its realisation.

The preservation of solutions as objects and conventions facilitates their (almost literal) incorporation within our behaviour where they are available for transformation. Cultural objectification circumvents the infinite regress faced by a general learner, by replacing its endosomatic isolation with a dual system in which exosomatic cultural knowledge is a memory, representation and specification of what and how an individual should learn.

The problem of what to learn is however faced by cultures at another scale and focus. Cultures may follow inconsistent beliefs and policies, can fail to learn, or learn the wrong lesson (Tuchman 1984; Bahn and Flenley 1992). This is not surprising considering the balance that culture has to maintain between conservation and innovation.

Is what culture learns purely operational? Can the development of culture be viewed in the same selective and adaptive light as evolution, mediated by memes (Dawkins 1982) rather than genes? Do technology and culture serve as a link between the intrinsic capacities of human cognition and a teleological frame in which human experience cognizes information that is significant not just for its survival but which allows it access to the laws of nature?

The significance of learning is perhaps most striking when we consider questions that we do not seem to be able to answer. As a species we are continually questioning reality and our experience. Yet there are some kinds of question we seem incapable of answering. For example, Idealism and Empiricism continue to reinvent their positions historically (Nativism, Realism) but now seem more akin to psychological attitudes rather than rationally conclusive positions. This can be interpreted in different, not mutually exclusive ways. The continual restatement of epistemological questions concerning the status of knowledge and how it arises indicates that such questions are important to us; but our ongoing inability to provide final agreed answers indicates that our cognition may not be very good at abstract thought or these teleological, cosmological questions may not be ultimately decidable.

Another pragmatic aspect of such abstract concepts is that like the operationalism of the technological thought they encapsulate and rationalise, they may be, of necessity, an open system. Perhaps without the overarching incompleteness and paradoxical nature of abstract thought we would not be able to operate as successfully at a pragmatic and technological level and so would not be able to innovate and learn as we do? For example, the ideas that the universe is finite or infinite are equally paradoxical. We find it difficult to conceive of space that is not bounded yet the scale of the universe seems to preclude this. The medieval view of the world being within a closed cabinet with the stars as windows allowing the light of heaven to shine in bears comparison to the contemporary idea that the universe itself is a form of simulation within another universe (Stonier 1997) or more humorously, an experiment to define the ultimate question (Adams 1979). All share the notion of the reality we know being in some way 'staged' within a context we can never know or understand.

Scientific history indicates that technology and culture is a systematic attempt at a best overall fit between concepts and observation (Pacey 1990). Observations or theories that don't fit a current paradigm are often discounted or ignored, e.g., the particle *vs.* wave theories of light (Kuhn 1962) or sun *vs.* earth-centred cosmology. Periods of conservation alternate with transformation. Culture switches between static and dynamic modes and vice versa, at the drop of an environmental/historical hat. The notion that there is final understanding of the 'world' to be arrived at; and furthermore the apprehension that we might be approaching it, would preclude the introduction and acceptance of insights at variance with this paradigm. This has occurred historically. The effect of teleological orthodoxy is well illustrated by the response of the Medieval Papacy (Tuchman 1984) to the empirically informed impetus of medieval science (Gimpel 1976). If we believe that we 'understand', this seems to undermine our impetus to find a different, possibly more truthful account of the world.

Thus it may be the case that humanity will continue to learn simply because our higher order cognition cannot come to a final conclusion about the nature of the world, or our understanding of it. While technology, the insights it invokes and the changes it initiates may innovate, there will always be an irresolvable distance between human thought and the reality it tries to comprehend and this will convince us that the nature of our cognition has not and cannot change.

## 4. How and what we think: Is cognition unbounded?

If cultural interaction and memory is accepted as a significant element in cognition, it raises the question of whether cultural cognition and technology are an unbounded system. Human cultural systems are open in the sense that they interact with their environment and our biosphere while relatively 'sealed' is not completely isolated (Swenson 1992). The question is the extent to which human cognition as a cultural system is constrained or bounded in what it can understand and conceive.

This question re-emphasises the relationship between how we think and what we think about, and the distinction we make between cognition as a system and the thought that arises through its actuation. If we assert that technology can change the nature of cognition, are we referring to the nature of the process of cognition or the nature of the world in the ideas, objects and processes that cognition manifests? Does the capacity to address different aspects of the world imply that cognition has changed irrespective of the biological organism in which it is actuated?

Evolution is a lawful process that realises systemic properties of physical form, experimenting with organisms that are viable in time and space. In one sense, the evolution of intelligence presupposes that it is immanent, potential or inherent in the physical world. This is a strongly anti-evolutionary view (Muller and Newman 2003). Evolutionary theory explains how something that did not exist gradually emerges or becomes. But this idea of emergence is rather equivocal. If some state (or thing) is 'actual'; then before it became so, it was possible, and if it was possible then its becoming is a realisation of that inherent potential. This potential is only bounded in the sense that it must conform to the laws of physics. If cognition is immanent in the physical world this implies that it precedes its cause, i.e., its genetic specification. It can only exist through processes describable through the laws of physics, yet these are themselves 'discovered' by it. The laws of physics have largely been discovered through our desire to improve technology. Do we believe the laws of physics are real and truthful or operational fictions? In this context the distinction of the material from the mental is more a reflection of our operational focus than any conclusion about the nature of anything.

## 5. Technology, representation, meaning, and mechanism

Does the history of technology reflect the importance of the cognitive mechanism or the ideas and tools it creates and uses? Physically, technology involves transduction or extension of the organisation of energy, and this has contributed indirectly to the ever more abstract organisational properties of human societies and their reliance on different trophic energy sources (Wilkinson 1973). No less pragmatic, but less directly physical, abstract information and knowledge technology is coeval with psychological externalisation and the exosomatic objectification of thought in signs and symbols, including language, music, writing, mathematics, images, and technology. Writing, numbers, the abacus and the computer, are all tools that we think of as redefining cognition, because they re-present ideas or an aspect of reality that is a shared product or property of our collective thought.

The idea of re-presentation is currently pre-eminent in framing theories of cognition. Its core meaning is 'to depict'. It has several (non-political and legal) connotations and associations; including constancy of form and temporal extension; it stands for some 'thing' or 'idea' and as such has inherited much of the epistemological nexus of the 'sign' (Land 1974). It also has the connotation of transformation of one form into another and of mediation between identities. Its semantics focuses on states rather than processes, which has drawn criticism from views that emphasise the ecology of informational flux and responsive interactions (Gibson 1979). The very notion of re-presentation, while active in the sense of 'doing-again', reflects the tendency of our thought to objectify itself, and reflects the processes of internalisation and externalisation.

For a representation to represent something implies homology between it and what it represents and particularly structural equivalence that encapsulates what is systemically or causally significant for the phenomena represented.

The power of re-presentation is pre-eminent in the history of numbers. Moving from ten fingers to an abacus was a generalisation of base ten. The abacus' facilitation lies in the memory inherent in the physical rows of beads representing powers of ten. It replaces mental arithmetic that we find hard, with an action we find easy, i.e., flicking beads about. The abacus frees memory, because it remembers in its structure. Arguably all significant mental representations have analogous effects; in the case of numbers by changing how the brain engages in numeric operations (Butterworth 1999), e.g., Arabic numerals facilitated place arithmetic (Flegg 1984).

Tools for thought re-present (transform) problems in two directions. Its users understand the representation and it reflects the reality or problem they direct it at. Thus re-presentation is a means of cognitive specialisation: a way of changing the structure of a problem so that our cognition can deal with it. If this is true it implies that a problem may be represented inappropriately. And this is the case. For example, while people experience difficulty with logical tests such as the Wason Test represented formally (using cards), when encapsulated in a social situation people find them much easier (Gigerenzer and Hug 1992).

Another example of representational relativity is Archimedes' geometric formulation of the calculus two thousand years before its algebraic 'discovery' (Netz 1999). The underlying problem Archimedes addressed is the same, and arguably the only difference is between the two ways of re-presenting the continuous world in discrete 'infinitesimals'. Re-presentation seems to be a technology for the transformation of informational problems so that they become more amenable to our intelligence, just as machines transform physical problems to be susceptible to our limited physical strength. Representational relativity is also reflected in dormant ideas. For example, the invention of the computer transformed the significance of Boolean Logic. It is almost as if, as Plato proposed, cognition finds what is already there, not necessarily driven by experience or functional necessity; but by rediscovering or adapting elements of an intrinsic mental repertoire. Another trivial example is the contractions used in telegrams and *txt msgs*. This form of shortening seems novel and technology specific, yet it has been used, probably for millennia, in surrogate languages, e.g. African drum languages (Herzog 1945).

The example of Archimedes answers the question of whether cognition is culture (tools and ideas) or the mechanism of the brain, by indicating that the reality is more complex. It shows how close their relationship is and also how distinct they are. Other species share with us some numeric skills; but through external representation our arithmetic skills have become transformed into the open system of mathematics.

The emphasis of our understanding of thought and its relationship to the world until recently was on ideas and their relationship to the world. Ideas 'mean' something we judge, accept or reject. With the Renaissance and Enlightenment interest started to shift from ideas to the mechanism that produces ideas. However, trying to decide which is significant produces a dilemma. If the mechanism is normative, given the historic relativity of ideas and concepts, the truthfulness of particular ideas is arguably inconsequential. On the other hand if ideas are paramount we will never understand the mechanism that underlies them because our view of it will always change. As Feyerabend (1999) has pointed out, the scientific method is itself a belief system based on a generalisation of the principles of evidence and conclusion in civil law. It is pragmatic but seemingly as *ad hoc* and imperfect as any incantation in discerning final causes. Science and the operational epistemology it promulgates is theoretical technology.

The strength of belief systems and construction of the world compounds the difficulty of separating what is intrinsic to brain, received from culture, and experienced in reality. In order to try and understand what is intrinsic to cognition *qua* mechanism we need to prise them apart; however, it is very difficult (some would say impossible) to step outside culture, for a culture constructs reality in ways that its members consider to be natural.

An example that illustrates the constructive nature of culture and science is the perception of tonality in music, which dominates western musical practice. Historically ideas of tonality within musical practice metamorphosed into rationalisations about how we perceive music (Meyer 1956). Tonality shifted from a *modus operandi* into an analytic construct and finally to a putative psychological mechanism. Yet attempts to ground tonality in psychoacoustics (Kameoka and Kuriyagawa 1969) or pitch structure (static context-independent schemas) and pitch function (dynamic time-dependent processes) (Butler 1989; Krumhansl 1983) have proved either inconsistent (Huron 2002) or inconclusive. Recent experiments (Houvinen 2002) indicate that tonality is not a correlative of a normative stimulus structure extracted by a mechanism 'tuned' to this 'intrinsic organisation'. Our perception is piecemeal. If presented with tonal structure we accommodate it. Thus, tonality is a stimulus organisation that can be recovered imperfectly in a number of ways from music that has been systematically (i.e., culturally) organised in terms of its theoretical or aesthetic (i.e., cultural) principles.

Representation is used by culture to construct experience and it is through this coherent mediation that individuals cognize. To understand more of the relationship between mechanism and thought we need to consider biological and physical views of culture and cognition that recognise cognition as a systemic property rather than as a mechanism.

## 6. Social cognition, tools, and stigmergetic process

The idea that cognition is not bounded by endogenous specification is reflected in theories of Situated Action (Suchman 1987), Distributed Cognition (Hutchins 1995), Activity Theory (Kaptelinin et al 1995), and biological theories such as Stigmery (Theraulaz and Bonabeau 1999). In all these, tools play a significant role in the mediation or structural and communicative aspects of collective action.

While other species possess aspects of human culture, there is no other species that possesses all its characteristics. The problem is to explain its characteristics without appealing to tautological pre-figurement. Answers include the replacement of genetic specification of behaviour with memes (Dawkins 1982) or generalisation of Machiavellian intelligence (Humphrey 1976). Another equally possible origin may be the development of the relationship between intrinsic endogenous models of the world and systems for communicating them. In this view culture as a system emerged and is evolving because of the ontological leeway that exists between culture (externalisations) and individual behaviour. The aspect of culture that reflects this most strongly is technology. Because it is concrete it exemplifies objectivity and because it is made it can be changed and improved. Technology is the open 'stigmergetic' property that allows the 'phenotypic' (structural) evolution of culture and the increments of understanding its objectifications facilitates. Theoretical explanations and ideas don't have to be truthful, or even to exist for behaviour to be efficacious.

For example, the mechanical principle of the lever has been used since the Upper Palaeolithic in the form of the spear thrower. This pragmatic exploitation continued for millennia before any scientific attempt to explicate the principles involved. Our understanding of levers developing from their use exemplifies the accretive transformational aspect of culture that is comparable to the biological principle of stigmery, where environmental modification induces behavioural changes in individuals.

A more recent example is Carnot's and successors' development of the laws of thermodynamics. Engineers and industrialists wanted to make the practical technology of the steam engine as efficient as possible. This led to understanding the relationship between an engine and its heat sink, which was then generalised into the 2nd Law of Thermodynamics. Thus, our view of life and the laws of physics are technologically inspired and informed, just as our view of organism as mechanism is.

How are we to understand stigmery in relation to human culture? Stigmery is mediated in much simpler ways than human culture and is apparently static. The organisation that imbues the anthill is conservative. However, in one essential respect stigmergetic activity is the basis for culture. The result of behaviour is artificial 'objects' that other individuals respond to. Stigmery and culture share this, and arguably reflect a single underlying organisational principle: that behavioural change can be self-induced by changing the environment, because it is the environment that 'specifies' behaviour. If the ability to behaviourally respond (as well as adapt) to change is general, i.e., involves learning, the stigmergetic principle develops into the altogether more powerful organisational principle of culture which is not only conservative but implicitly innovative.

The emergence of culture from an ancestral stigmergetic form was gradual, and until the last cultural stage the evidence is that it was phenotypically specific. The steps from stigmery to culture may have included stimulus enhancement and response facilitation (Mithen 1996). While stigmery involves self-inducement of behaviour through environmental manipulation, stimulus enhancement involves a heightened focus on environmental cues from another organism's behaviour. Response facilitation occurs where activity stimulates another organism to mimic the act. From these modes of response it is possible to conceive of a step to intentional imitation and on to general learning. The crucial property that general learning facilitated was our ability to respond to complex and arbitrary behavioural cues and objects. As culture developed it replaced the phenotype as the specifier of behaviour, multiplying enormously the options for behavioural, organisational and informational complexity (Tomasello 1999).

The role of behaviour in changing and structuring the environment is also core to Activity Theory, which stresses the importance of artefacts, and their mediating function at the juncture of consciousness and behaviour. Behaviour is construed in Activity Theory as a hierarchical process of activity, actions and operations that have corollaries in the hierarchy of motives, goals and conditions through which activity takes place.

Activity Theory emphasises the dynamic of behaviour rather than the objective stasis of the representations and conceptions that emerge from it and reflect it. In it artefacts are conceived to mediate an ongoing dialectic of intention and conditional adjustment. Tools are considered to be intrinsically contextual and consequently lend themselves to hierarchical definition of activity. Activity Theory is like the obverse of stigmery, arguing that repeated activity perpetuates its own identity and becomes objectified. Activity Theory is recursive in the sense that it proposes a dialectic between action and object; abstractive because complex actions are embedded within and shared between activities, and viewed as categories (objects) of action. It is iterative in the sense that similar activities are re-enacted, and historic in its directional change.

The conjunction of abstraction and iteration in objectified activity is what leads to the accretion of both physical and abstract artefacts. However, the process of accretion is not simply accumulative, but transformative. The tools that are transformed include signs and symbols, written language, scientific notations and practices: in fact all forms of conventional organisation of behaviours are susceptible, as well as tools and machines for doing things. For Vygotsky (1986) the transformation of activity through artefacts reflects the linkage between organism and world: this is essentially stigmergetic.

The interaction of technology, actions and ideas is observable directly within the formulation of activity such as playing the piano (Sudnow 2001). The composer Ligeti (1997: 8–9) wrote about composing for the piano:

> I lay my ten fingers on the keyboard and imagine music. My fingers copy this mental image …, but this copy is very inexact: a feedback emerges between idea and …. execution. This feedback loop repeats itself …. a millwheel turns between my inner ear, my fingers and the marks on the paper. The result sounds completely different from my initial conceptions: the anatomical reality of my hands and the configuration of the piano keyboard have transformed my imaginary constructs. ….. The criteria are only partly determined in my imagination; …. they also lie in the nature of the piano — I have to feel them out with my hand.

Ligeti's description of technology imprinting itself on activity reflects 'technology in action'. But we also need a wider account of the adaptive emergence of culture. The theories of Autokatakinesis and Autopoeisis (Swenson 1992; Maturana and Varela 1980) are both focused on systemic properties and how these are reflected in the ongoing flux of adaptation and order observable through evolution.

## 7. Autokatakinesis, autopoeisis: Information and technology

Autopoiesis is a theory that distinguishes animate from inanimate by defining a living organism in terms of functional organisational unity that survives the physical replacement of its constituents. Autopoiesis is not a theory of final principles or causes but a definition of a process. Autokatakinesis on the other hand is a teleological theory. It holds that life is an expression of the 2nd Law of Thermodynamics: the imperative of entropy through which the energy of the universe becomes evenly distributed. The obverse of energy is order; and life, in creating ever more complex orders of organisation is a means through which the universe dissipates energy as rapidly as possible (Swenson and Turvey 1991). Culture and technology are forms of organisation that facilitate this process.

Autokatakinesis rationalises 'why' life emerged. Autopoiesis describes how it is manifest in a particular form of physical order, i.e., the consumption of energy through chemical conversion. The dividing line between behaviour and culture is of an analogous order to that between the physical and the biological (the inanimate and animate). The separation of inanimate from animate lies in growth through accretion or intussusception. The emergence of life from pre-biotic may have involved proto-biotic systems that involved both (Stonier 1992). Similarly, the evolution of multi-cellular forms was an increment in the endosomatic organisational complexity of the phenotype. And, as the 'embedding' of unicellular forms produced complex forms and behaviour, so organisation and encapsulation of the information across single 'cells' of the phenotype, results in multi-cellular culture. Its 'memetic' rather than genetic specification is manifested in the openness of the relationship between the mechanism of the phenotype and culturally specified ideas, practices and tools (Dawkins 1982).

The inherent instability of (biological) systems is a corollary of emergence. Autokatakinesis suggests that the emergence of higher orders of organisation is inevitable, providing a rationale that cognition can and will change and that the change involved will lead to a redefinition of its nature through cultural evolution.

The rationalisation of the distinction of life from the inanimate in Autopoiesis finds an echo in the integrity or unity of culture. While recognising that analogies between society and the functionality of a body must be circumspect, the functional unity of culture is striking. Culture, while encapsulating apparently *ad hoc* historically derived responses and practice, as well as being capable of adjustment and adaptation, is highly susceptible to the disruption of its elements. Few cultures have recognised this quite as cogently as the Japanese who expelled missionaries and banned the gun between the 16th and 19th centuries because of their social and political effects (Perrin 1979). The history of many cultures worldwide since the spread of European technology reflects the profound influence that changes in technology have. The rupture of cultural integrity can lead to 'millenarian' responses that vary from the explicitly political to the religious (Wilkinson 1973; Burridge 1969).

Autokatakinesis is a strong rationalisation of Ecological Realism's (also Direct Realism and Ecological Psychology) theories of perception and action (Gibson 1979). These are congruent with Activity Theory, as well as with Situated and Distributed Intelligence. Ecological Realism emphasises the importance of informed activity and the evolution of perceptual and action systems to maximise the ability of an organism to extract higher-order specifications of energy and their environmental sources. For Ecological Realism human cognition is not defined by the objects of its thought, but by its structure as activity and it is this activity that consumes or dissipates energy.

Autokatakinesis and Autopoiesis reflect technologically originated ideas. Ecological Realism, hard-core computational reductionism and distributed theories of cognition are all 'information centred' theories. It has been proposed that information is as fundamental an aspect of the universe as energy and matter (Stonier 1992, 1997). The Autokatakinetic view is that information reflects organisation that in turn reflects thermodynamic improbability and the 'work' needed to create it. Human cognition is an ever more complex organisation of information. The historic development of human cultures and their coherence reflects the physical concomitance of information and organisation, and the fact that when the organisational 'bonds' of ideas, rules, practice, and laws are loosened through neglect, decline, revolution, environmental change or natural disaster, the integral, incremental accretion of organisation complexity is compromised. Human history has frequently exemplified this. The history of technology may also indicate that culture itself was something that our ancestors learned.

## 8. The timescale and rate of technological change

While technology is not unique to humanity no other animal makes composite tools or applies them as indirectly. Our tools and machines have developed dramatically in every area: transport, energy utilisation, agriculture, and manufacturing (Ingold 1993). In evolutionary timescales the human story is very short. The rate of technological change has been uneven, but has become progressively more rapid and its global ramifications more integrated (Pacey 1990). The development of culture and technology indicates a piecemeal systemic evolution in which complex organisation is intrinsically unstable (Tainter 1988). A defining theme of world history has been that civilizations rise and fall and societies pass through revolutionary phase shifts in organisational complexity and organisation: the Upper Palaeolithic Explosion, the Neolithic Revolution, Urbanisation, the Industrial Revolution, and Globalisation. Technology incrementally extends past inventions (Heilbroner 2001).

The emergence of global civilization started in several local areas of the Middle East, India, China and the Americas and spread regionally. Even though European, African and Asian Empires traded, the links for a long time were tenuous. Regional empires continued in all core areas, until the first wave of globalisation in the Spanish and Portuguese Empires. The next step towards globalisation in the 17th to 19th centuries involved cultural changes in financial mechanisms, industrial technology and communications infrastructures (railways, steamships, telegraph) that linked the globe together in an unprecedented way; although all previous civilizations also reflect that new orders of organisation have been predicated on technology and organisational objectification, including communication systems, systems of law, written records, economic centralisation, machines, etc. There seems no reason why future re-organisation, such as mediated by the World Wide Web will not repeat these earlier changes.

However, the activity that underlies tool use and manufacture is similar in its overall form across cultures and throughout history. What differs is the complexity of component function and of their organisation in manufacture and use.

For example, an early form of indirect tool manufacture was the Levallois technique in which a prepared 'tortoise' shaped stone core determined the shape of a flake. Making a Levallois flake involves a series of planned actions in a behavioural hierarchy such as proposed by Activity Theory. A complex technology such as a car, telecommunications satellite or atomic bomb is made in a similar way, though their properties and consequently the activity is more complex. Ultimately all rely upon the temporal integration of actions. Even the scientific method: iterative hypothesis and experiment, can be seen as an abstraction of the processes involved in making a stone tool.

The developments that led from making something like a Levallois flake to an atomic bomb were piecemeal, step-wise accretions that relied on the memory carried in the tools themselves. The archaeological record is a mass of these incremental typological series. Natural and technological evolution are similar in their adaptive accretion and refinement through differentiation and specialisation of parts and attributes. Each increment is a change born from the previous state, upon which it depends. Just as evolution did not create the human eye *ab initio*, so humanity did not make the atomic bomb in the Stone Age. How appropriate a tool is for its purpose determines its selection for improvement. Evolution likewise involves selection, albeit over very long periods of time. Also, although incremental and accretive, tools share with biological adaptations an aspect of serendipitous or opportunistic pre-figuring (Gould and Lewontin 1979). Examples include the use of gunpowder for fireworks, then as a weapon; or the development of the World Wide Web from Arpanet.

In a way comparable to the evolution of organs, technology involves informational and organisational increments in constancy, complexity and immediacy (writing, printing, telecommunication) that in conjunction have reformulated our responses to the world. Technology in this sense is a dynamic or temporally extended form of stigmery. In considering the contribution of biology and technology (culture) to human cognition there is perhaps a useful insight to be gathered from considering the relationship of technological change to the culture and emergence of our own species.

Tools and technologically defined spaces have existed at least since Australopithecines erected wind breaks at their campsites in the East African rift (Leakey 1971). Early technologies, from the evidence, were quite static. Lower Palaeolithic assemblages changed only in details over ≈1.5 million years. Oldowan and Acheulian are associated with the hominid species of Ergaster and Erectus. The development of Middle Palaeolithic/Stone Age technologies around 120k BC is linked with early forms of modern Homo Sapiens in Africa, Europe and Asia. In Europe the Neanderthals are found with Middle Palaeolithic (Mousterian) assemblages (Klein 1999; Mithen 1996). Although the evidence is incomplete, the prior technological stasis and the synchronisation of technological change with the appearance of new species, indicates that for these ancestral hominids technical skill and by implication cognition was still largely phenotypically specified.

Albeit sophisticated in one sense, Middle Stone Age Technology is not indicative of any capacity to consider the process of its fabrication (or for that matter any other tool) separately from the act of making it. We can distinguish between being able to a) memorise an action such that it can be repeated at arbitrary times through an act of will; b) abstract an action in a way that allows the act of planning to be transferred to other situations and c) the abstraction that allows the act of planning itself to be considered as an abstraction. Arguably, what underlies 'abstraction' is the ability to identify significant properties and recursively re-consider processes (actions) as states (objects) or patterns. In this way, just as embedding components leads to more complex machines, so more complex information arises from the structure of embedded information.

It can be argued that this process of abstraction underlies the transformation of tools for action into tools for thought. The capacity to learn (abstract and generalise) while facilitating experimentation and creativity also involves mistakes and impracticality, a technological equivalent of evolution's 'hopeful monsters'. Perhaps our ancestors started to make 'useless' tools that had no function, but that nevertheless reflected or signified their own production and this transduction of action into object was then available as the basis of signification and symbolic thought.

The transformation of tools from physical transducers to informational re-presenters is difficult to pinpoint. The argument is susceptible to tautology. The idea that a useless thing can be a sign presupposes that the perceiver has the ability to recognise its potential signification. So for an evolutionary account we require evidence of a process in which the significance or meaning of the act of creating signs was itself emergent or incremental and not associated with speciation.

Modern humans are thought to have evolved in Africa sometime before 100k BC. However, the leading-edge technology continued to be of Middle Stone Age type until *circa* 40–35k BC. Thus, after the appearance of modern humans there was a period of some 40–50k years in which we were confined to Africa and in which lithic technology changed little. Then, sometime after 50k BC technology and representation exploded. There is a period of rapid cultural replacement that was essentially an invasion by modern humans of the rest of the world (Klein 1999). The species resident in Europe, the Near East and Asia (Neanderthals and Erectus) were replaced. In Europe intermediate technologies such as the Chatelperronian associated with Neanderthal occupation are interleaved in some sites with Aurignacian blade technology and modern humans. Chatelperronian techniques probably indicate borrowing by the Neanderthals (Klein 1999).

There are aspects of this story that counter the simple alignment of cognition to species, and of our species' learning and innovation, arising solely from changes bound to neurological architecture and function. The emergence of modern humans around 100k BC is synchronous with the appearance of proto 'pressure blade' technology in North Africa (McBurney 1967), yet initially neither it nor its maker moved beyond Africa, nor is there widespread evidence of representation. Then around 40k BC pressure blade techniques diffused widely and art and representation appear, as well as musical instruments associated with modern humans (Klein 1999; Marshack 1972; Leroi-Gourhan 1968; Mithen 1996).

The emergence of representational systems such as paintings and artefacts that mediate or externalise imagination and reason is taken to be a fundamental aspect of our species' cognition (Mithen 1996). However, the de-synchronisation of the appearance of our species and culture raises the question of what exactly the 'explosion' of symbolic representation *circa* 40k BC indicates. Why is the emergence of the new species *circa* 120–100k BC not synchronised with the technological and representational revolution? One explanation is that our symbolic, culturally mediated thought developed during this period.

The appearance of Cave Art *circa* 40–30k BC is not the first eruption of representation. The earliest example of such abstraction is around 70k BC, 30k years before the Upper Palaeolithic 'explosion' in Europe. It is a piece of inscribed ochre from the Blombos cave in South Africa (Henshilwood et al. 2002). The inscription is a hatched geometric design like a series of overlapping XXXs surrounded by a border. The associated lithic culture is Middle Stone Age.

The act of making this design or pattern reflects not only indirection (the fabrication of an engraving tool has preceded it), nor is it just the transference of an action from one domain to another (such as from flaking a stone to sharpening a stick). The Blombos ochre seems to involve the explicit transformation of an action into an object. The engraved pattern states or 'represents' in its static organisation an objectification of the act that produced it. If it is an abstraction of action directed at nothing but its own creation it is open to represent anything to its maker, and it is difficult to see in the Blombos engraving anything other than the objectification of its own creation. The supposition here is that such an object meant something to its maker because it was in other respects useless. Even at the distance of 70k years the observer knows this object meant something. Our imagination recognises the intention of signification. Our interpretation engages in the same projective, imaginative activity as its maker and users, except we cannot grasp its meaning because we do not share their culture.

The creation of 'useless' artefacts such as the Blombos ochre engraving is arguably seminal for human cognition. The creation of empty, useless objects that signify their own creation, amounts to producing objectifications through which we understand our own cognitive activity. Because once we have externalised them this is what we can do with them, namely, to consider how we have made them and what they mean. We don't cut down trees or hunt animals with them; we think with them. The Blombos engraving is geometric, made by a series of strokes. In terms of representational development it is interesting that the ordered sequence of marks/stokes is iterative, indicating a physical or manipulative algorithm.

If the human capacity to learn is strictly phenotypic, why did it take 40–50k years for modern humans to colonise the world with Upper Palaeolithic technology and art? What does the Upper Palaeolithic ramp imply? The timescales and coordination of the emergence of technologies and representations in the Upper Palaeolithic can be interpreted as indicating that the cognition of modern humans evolved through culture. Could it be that 'last-minute' phenotypic adaptation was selected for because it enabled cultural learning and conservation? If our learning and culture arises simply from the intrinsic properties of our cognitive hardware why didn't Upper Palaeolithic Culture appear worldwide at the same time? Why has the rate of technological development varied geographically and historically?

If human cognition including general learning is simply phenotypic, the cultural stasis from *circa* 110k to 40k BC is inconsistent: but not if technology itself played a key systemic role in controlling and facilitating cultural cognitive development and learning. The evidence of the pre-historic record is certainly consistent with an interpretation of Culture itself emerging through such an informational, representational and technological bootstrapping effect and that cultural cognition is not a normative behavioural protocol that simply mediates intrinsic capacities (Barkow et al. 1992; Mithen 1996). Rather it was an emergent phenomenon that became progressively independent of its phenotypic hardware and more powerfully specified in its organisation of information. Arguably, culture is a behaviour that our species had to learn. In this view our species developed culture because our phenotype provided the general learning capacity that facilitated the transformation of 'dynamic-stigmery' or 'proto-culture' into culture. The development of learning and culture initially progressed at a very slow rate. Tools, concepts and ideas were conservative. It took some 40k years of development to arrive at its explosive 'appearance'. Until this the idea of memes conserving behaviour seems apposite; however, once culture took off, the notion of meme is arguably too static to capture the integrative dynamic of cultural conservation and innovation in organisation, ideas and tools. Since culture became dominant it has demonstrated its independence from the phenotypic intelligence of the individual by historically switching between conservation and innovation of information as circumstances have allowed. When the opportunity presents itself it moves towards greater complexity and informational integration.

The developments from 100k to 40k BC commenced a process of cultural accumulation that continues to accelerate in informational complexity. The Upper Palaeolithic 'cultural revolution', manifested in cave and parietal art (Leroi-Gourhan 1968), ended abruptly with the last Ice Age around 12k BC. During its latter stages, after 16k BC, other re-presentational 'abstract' technologies appeared, including marked bones indicating an early form of calendar (Marshack 1972). Writing appeared in the Near East some time around 4–5k years later. By late antiquity, Greek civilization made mechanisms such as the Antikytheran Mechanism for calculating heavenly events (de Solla Price 1959). After another 1.5k years, in the 19th century Babbage designed the Difference and Analytical Engines, and after the merest blink of a cosmological eye, Turing defined the computer and Von Neumann realised its digital electronic manifestation. Subsequently the merging of digital communications and computers has created the World Wide Web, which is the largest informational integration the world has yet seen.

## 9. Conclusions: Technology, nativism and realism

This essay commenced by asking whether technology changes cognition. It asserted that cognition is not usefully seen as confined to the 'computations' of individual brains, but as the systematic functional organisation of individuals' thought via signification and artefacts. This distinguishes the endogenously specified function of the mechanism from the exogenous organisation of culture.

The cultural cognitive mechanism is manifest in the representations, technologies, practices, ideas, perceptions, consciousness and sensations that are capable of manifestation within a re-cognizer, and this is part of the intrinsic specification of our species' capacity. This hardware specification is necessary but not sufficient for human cognitive organisation. To be fully functional the relationship between it and culture has to be complementary and iterative. To characterise it as stigmery fails to account for the transforming history of technology and culture. Culture mediates learning and conservation through the integration of internalised and externalised abstractions of behaviour. These underpin the degrees of complexity in the composition of tools and representations, historic variability and cultural change.

The first section argued against the computational metaphor as an adequate model of cognition, because it provides no insight into the historic development of culture. We attempted to reformulate our original question by suggesting that the ideas and actions that cognition instantiates are not a closed set; that ideas are learned and change, and it is ideas, mediating between mechanism, behaviour and the world, which *de facto* constitute operational cognition.

The transformation of re-presentations and the facilitating effect of reorganisation suggest that representation is a form of specialisation that mediates between intrinsic capacities and specific problems. This interaction is affected by how well a tool fulfils a function, the ideas associated with it, how amenable it is to development, and cultural context. It is pragmatic rather than teleological, evolutionary as well as intentional.

The ultimate abstraction and externalisation of technology's operational utility is science. In science as theoretical technology we see the strongest indication of the duality of cognition as a combination of phenotype specification and technology. However, recognising the extension of human cognition beyond its phenotypic specification also reveals that openness involves indeterminacy.

Mistakes are part of learning and this implies that the human cognitive cultural system is open. The separation of mentality from brain in cultural views of cognition is a form of dualism. While mind is co-terminal with the material processes of brain activity, the ideas it can actuate are not specified, but determined by the meaning inherent in the informational organisation it instantiates. Historically, existentially, systemically and experientially it is the nature of the ideas that the mind/brain absorbs and actuates that have been significant.

The difficulties that arise from trying to understand the relationship between mechanism and its experience are evident in the view of intelligence developed within cognitive theory since the 1950s. Cognitive science rejected dualism; embraced the empirical (materialist) imperative of reducing mind to brain and an idealist/rationalist view of the principles of intelligence as independent of the brain. It emphasised universality of function as the defining property of the mechanism of intelligence and reduced culture's role to developmental tuning. However, this view does not explain historic development and also presents a problem in determining what and where meaning arises within such a system.

One resolution to this conundrum is a Direct Realist reconciliation of Nativism and Empiricism in a form of evolved Idealism, in which human cognition is an emergent process that transforms and is transformed by information. It is founded on an evolved informational grounding, reflected in 'perceptual invariance' and developed through the dualism of brain and culture. In this view culture constructs cognitive organisation through conservation and control of learning. It is intrinsically indeterminate, error prone and evolutionary, but open. The theories of Autopoiesis and Autokatakinetics provide biological and physical rationales for this account of the transformative reorganisation of information through culture and for the organisation of information in representations and artefacts being closely linked to the emergence of open learning.

The juxtaposition of mechanism and cognition is well illustrated by Chomsky's account of language acquisition (Chomsky 1988). Chomsky's view of language structure is paradigmatic for cognitive science in seeking functional specification in structural encapsulation. It recasts prior explanations of the variety of extant languages as developments from a single ancestral language (Eco 1999; Ruhlen 1994), in terms of the coextensive surface variation of superficial grammars and an underlying constancy of deep structure.

But does deep structure convincingly explain the multiplicity of languages? If, as deep structure implies, differences between languages are superficial, where does meaning reside? If deep structure is 'real' then arguably linguistic concepts have no existential correlative. We think, but we don't know what we think, because we are confined to the superficiality of language.

What is the advantage of diverse languages (and for that matter technologies, representations and ideas) if they are all in 'reality' the same? What is the advantage of a transformation between superficial and deep structure? From an evolutionary viewpoint there should be an adaptive advantage in the diverse instantiations of language — in which case language differences are significant within a cultural system. In this view the generality of the phenotypic learning mechanism, has adapted to understand cause and effect rather than to reconcile our phenotypically specified cognitive deep structure to the world. This generality is the main source of openness in culture and technology as well facilitating the imprinting of culture on the phenotype. Pragmatically, existentially and historically it is the cultural system that defines cognition.

Although the evidence presented here is inconclusive, it suggests that one coherent characterisation of cognition is as a systemic property involving individual minds/brains and culture in a dualistic and information transforming relationship. Autokatakinesis and Autopoiesis provide biological rationales for the emergence of order in information and its perpetuation. In this, technology and representation have played leading roles in conservation and innovation. Arguably it is the capacity of culture historically to switch between and utilise these conservative and innovative modes that indicates most strongly that human cognition is not phenotypically constrained and that human technology and cognition form an open and potentially unbounded system.

## Notes

1. Please see Section 6 for an outline of Stigmery.

2. According to Johnson Laird (1983) Craikian automata are an intermediate form. They lie between Cartesian automata that are non-symbolic and have no awareness, i.e., completely automatic reactive entities; and self-reflective automata that possess the capacity to embed models recursively and hold models of their own modelling capacity, thereby facilitating intentional behaviour.

## References

Adams, R. 1979. *The Hitch Hiker's Guide to the Galaxy*. London: Heinemann.
Bahn, P. and Flenley, J. 1992. *Easter Island Earth Island*. London: Thames and Hudson.
Barkow, J.H., Cosmides, L., and Tooby, J. 1992. *The Adapted Mind*. Oxford: Oxford University Press.
Burridge, K. 1969. *New Heaven, New Earth: A Study of Millenarian Activities*. Oxford: Basil Blackwell.
Butterworth, B. 1999. *The Mathematical Brain*. London: Macmillan.
Butler, D. 1989. "Describing the perception of tonality in music: A critique of the tonal hierarchy theory and a proposal for a theory of intervallic rivalry". *Music Perception* 6(3): 219–242.
Chomsky, N. 1988. *Language and Problems of Knowledge*. Cambridge, MA: The MIT Press.
Dawkins, R. 1982. *The Extended Phenotype*. Oxford: Oxford University Press.
Dennett, D.C. 1991. *Consciousness Explained*. London: Penguin Books.
Eco, U. 1999. *Serendipities: Language and Lunacy*. New York: Weidenfeld and Nicholson.
Edelman, G.M. 1988. *Topobiology: An Introduction to Molecular Embryology*. New York: Basic Books.
Feyerabend, P. 1999. *The Conquest of Abundance*. Chicago: The University of Chicago Press.
Flegg, G. 1984. *Numbers: Their History and Meaning*. London: Penguin Books.
Fodor, J. 2001. *The Mind Doesn't Work That Way*. Cambridge, MA: The MIT Press.
Gibson, J.J. 1979. *The Ecological Approach to Visual Perception*. Hillsdale, NJ: Lawrence Erlbaum.
Gigerenzer, G. and Hug, K. 1992. "Domain-specific reasoning: Social contracts, cheating and perspective change". *Cognition* 42: 127–171.
Gimpel, J. 1976. *The Medieval Machine*. London: Victor Gollancz.
Goody, J. 1986. *The Logic of Writing and the Organisation of Society*. Cambridge: Cambridge University Press.
Gould, S.J. and Lewontin, R.C. 1979. "The spandrels of St. Marco and the Panglossian paradigm: A critique of the adaptationist programme". *Proceedings of the Royal Society of London* B 205(1161): 581–598.
Harnad, S. 1990. "The symbol grounding problem". *Physica* D 42: 335–346.
Heilbroner, R.L. 2001. "Do machines make History?". In M.R. Smith and L. Marx (eds), *Does Technology Drive History?*. Cambridge, MA: The MIT Press, 53–65.
Henshilwood, C.S., d'Errico, F., Yates, R., Jacobs, Z., Tribolo, C., Duller, G.A.T., Mercier, N., Sealy, J.C., Valladas, H., Watts, I., and Wintle, A.G. 2002. "Emergence of modern human behaviour: Middle Stone Age Engravings from South Africa". *Science* 295:1278–1280.
Herzog, G. 1945. "Drum-signalling in a West African tribe." *Word* 1: 217–238.

Huovinen, E. 2002. *Pitch-Class Constellations: Studies in the Perception of Tonal Centricity* (Acta Musicologica Fennica 23). Turku: Finnish Musicological Society.
Humphrey, N.K. 1976. "The social function of intellect". In P.P.G. Bateson and R.A. Hinde (eds), *Growing Points in Ethology*. Cambridge: Cambridge University Press, 303–318.
Huron, D. 2002. "A new theory of sensory dissonance". In *Proceedings of the 7th International Conference on Music Perception and Cognition*. Sydney, 2002, 273–276.
Hutchins, E. 1995. *Cognition in the Wild*. Cambridge, MA: The MIT Press.
Ingold, T. 1993. "Tool-use, sociality and intelligence". In K.R. Gibson and T. Ingold (eds), *Tools, Language and Cognition in Human Evolution*. Cambridge: Cambridge University Press, 429–446.
Johnson-Laird, A. 1983. *Mental Models*. Cambridge: Cambridge University Press.
Kameoka, A. and Kuriyagawa, M. 1969. "Consonance theory Part I: Consonance of complex tones and its computation method". *Journal of the Acoustical Society of America* 45(6): 1460–1469.
Kaptelinin, V., Kuutti, K., and Bannon, L. 1995. "Activity theory: Basic concepts and applications". In B. Blumenthal, J. Gornostaev, and C. Inger (eds), *Human Computer Interaction*. Berlin: Springer, 189–201.
Klein, R.G. 1999. *The Human Career: Human Biological and Cultural Origins*. Chicago: The University of Chicago Press.
Krumhansl, C.L. 1983. "Perceptual structures in tonal music." *Music Perception* 1: 28–62.
Kuhn, T. 1962. *The Structure of Scientific Revolutions*. Chicago: The University of Chicago Press.
Land, S.K. 1974. *From Signs to Propositions: The Concept of Form in Eighteenth Century Semantic Theory*. London: Longman.
Leakey, M.D. 1971. *Olduvai Gorge. Vol. 3: Excavations in Beds I and II, 1960–63*. Cambridge: Cambridge University Press.
Leroi-Gourhan, A. 1968. *The Art of Prehistoric Man in Western Europe*. London: Frances and Holland.
Leroi-Gourhan, A. 1993. *Gesture and Speech*. Cambridge, MA: The MIT Press.
Lienhardt, R.G. 1961. *Divinity and Experience: The Religion of the Dinka*. Oxford: Clarendon Press.
Ligeti, G. 1997. *Notes for Études pour piano*. In *György Ligeti Edition*. Sony, catalogue nr: SK62308.
McBurney, C.B.M. 1967. *The Haua Fteah (Cyrenaica)*. Cambridge: Cambridge University Press.
Marshack, A. 1972. *The Roots of Civilization: The Cognitive Beginnings of Man's First Art, Symbols and Notation*. New York: McGraw Hill.
Maturana, H. and Varela, F. 1980. *Autopoiesis and Cognition — The Realization of the Living*. Dordrecht: Reidel.
von Melchner, L., Pallas, S., and Sur, M. 2000. "Visual behaviour mediated by retinal projections directed to the auditory pathway". *Nature* 404: 871–876.
Meyer, L.B. 1956. *Emotion and Meaning in Music*. Chicago: The University of Chicago Press.
Mithen, S. 1996. *The Prehistory of the Mind*. London: Thames and Hudson.
Muller, G.B. and Newman, S.A. (eds) 2003. *Origination of Organismal Form*. Cambridge, MA: The MIT Press.

Netz, R. 1999. *The Shaping of Deduction in Greek Mathematics: A Study in Cognitive History*. Cambridge: Cambridge University Press.
Newell A. and Simon H.A. 1976. "Computer science as empirical inquiry: Symbols and search". *Communications of the Association of Computing Machinery* 19(3): 113–126.
Norman, D.A. 1987. *The Psychology of Everyday Things*. New York: Basic Books.
Pacey, A. 1990. *Technology in World Civilization*. Cambridge, MA: The MIT Press.
Pacey, A. 1999. *Meaning in Technology*. Cambridge, MA: The MIT Press.
Perrin, N. 1979. *Giving Up the Gun: Japan's Reversion to the Sword, 1543–1879*. Jaffrey: David Godine.
Putnam, H. 1991. *Representation and Reality*. Cambridge, MA: The MIT Press.
Rheingold, H. 2000. *Tools for Thought*. Cambridge, MA: The MIT Press.
Ramachandran, V. 2003. "Neuroscience — the new philosophy". *The Emerging Mind, The BBC Reith Lectures 2003, Lecture 5*. http://www.bbc.co.uk/radio4/reith2003/lecture5.shtml.
Ruhlen, M. 1994. *The Origin of Language: Tracing the Evolution of the Mother Tongue*. New York: John Wiley.
Searle, J.R. 1980. "Minds, brains and programs". *Behavioral and Brain Sciences* 3: 417–457.
de Solla Price, D.J. 1959. "An ancient Greek computer". *Scientific American* 200(60): 60–67.
Stonier, T. 1992. *Beyond Information: The Natural History of Intelligence*. Berlin: Springer.
Stonier, T. 1997. *Information and Meaning: An Evolutionary Perspective*. Berlin: Springer.
Suchman, L. 1987. *Plans and Situated Actions: The Problem of Human-Machine Communication*. Cambridge: Cambridge University Press.
Sudnow, D. 2001. *Ways of the Hand: A Rewritten Account*. Cambridge, MA: The MIT Press.
Swenson, R. 1992. "Autocatakinetics, yes — Autopoiesis, no: Steps toward a unified theory of evolutionary ordering". *International Journal of General Systems*. 21: 207–228.
Swenson, R. and Turvey, M.T. 1991. "Thermodynamic reasons for perception-action cycles". *Ecological Psychology* 3(4): 317–348.
Tainter, J.A. 1988. *The Collapse of Complex Societies*. Cambridge: Cambridge University Press.
Tomasello, M. 1999. *The Cultural Origins of Human Cognition*. Cambridge, MA: Harvard University Press.
Theraulaz, G. and Bonabeau, E. 1999. "A brief history of stigmery". *Artificial Life* 5: 97–116.
Thomson D.F. 1964. "Some wood and stone implements of the Bindibu Tribe of Central Western Australia". *Proceedings of the Prehistoric Society* 30: 400–422.
Turing, A.M. 1950. "Computing machinery and intelligence". *Mind* 59(236): 430–466.
Tuchman, B. 1984. *The March of Folly*. London: Michael Joseph.
Vygotsky, L. 1978. *Mind in Society: The Development of Higher Psychological Processes*. Cambridge, MA: Harvard University Press.
Vygotsky, L. 1986. *Thought and Language*. Cambridge, MA: The MIT Press.
Wilkinson, R.G. 1973. *Poverty and Progress*. London: Methuen.

# Radical Empiricism, Empirical Modelling and the nature of knowing*

Meurig Beynon
University of Warwick

This paper explores connections between Radical Empiricism (RE), a philosophic attitude developed by William James at the beginning of the 20th century, and Empirical Modelling (EM), an approach to computer-based modelling that has been developed by the author and his collaborators over a number of years. It focuses in particular on how both RE and EM promote a perspective on the nature of knowing that is radically different from that typically invoked in contemporary approaches to knowledge representation in computing. This is illustrated in detail with reference to the modelling of several scenarios of lift use. Some potential implications for knowledge management are briefly reviewed.

This paper considers the potential significance of William James's philosophic attitude of 'Radical Empiricism' (RE) (James 1996) in relation to contemporary problems of knowledge representation in the information sciences. Current trends in computer technology and use provide a strong motivation for reviewing RE in this light. For instance, as Gooding (1990) observes, our understanding of how scientific knowledge relates to interaction with the natural world and with our peers is challenged by the development of virtual reality environments, and the role that virtual experiments have come to play in science. Such considerations have prompted a reappraisal of the fundamental assumptions that inform the logicist approach to knowledge representation in AI, and called into question the extent to which knowledge is mediated by language rather than engagement with the world (cf. Cantwell-Smith 2002; Turner 1996). In this connection, the relevance of RE stems from the priority it ascribes to 'pure experience', and its contention that (to paraphrase William James) the whole of the nature of knowing can be put into experiential terms (James 1996: 56). The problematic aspect of RE, as identified by Bird (1986), is that it is of its essence inarticulate: "… James's pure experience has to be such that nothing can be said about it, if it is to fulfil the role for which it is cast". This distances RE both from the mainstream philosophical traditions, and from the received views of computer programming as intrinsically bound up with formal languages and logical specification.

Empirical Modelling (EM) is an approach to computer-based modelling that has been developed by the author and his collaborators at the University of Warwick over several years (for background, see Beynon 1999, 1997; Beynon et al. 2001; Beynon and Sun 1999; Beynon and Russ 1991; Beynon, Rungrattanaubol, and Sinclair 2000; Beynon, Rasmequan, and Russ 2000; the EM website; the EM archive). The product of an EM exercise is first and foremost to be regarded as a source of experience whose interpretation by the modeller is not preconceived, but is to be established in the mind of the modeller through an association between experience of the model and experience external to the model. Knowledge in such a context has the qualities that James (1996) attributes to knowledge: it is a personal awareness on the part of the modeller that one experience stands in a particular relation to another. To borrow James's expression, experience of interaction with the model is 'an experience that knows another' that can act as a substitute for experience external to the model in a definite practical sense (James 1996: 61).

This paper explores the extent to which, through model-building using EM, it is possible to track James's exposition of the empirical roots of knowledge, with its emphasis on the fundamental significance in sense-making of our capacity to experience conjunctive relations between things. This exploration touches on many issues topical in modern computing that are addressed in James's account of RE, such as the nature of consciousness (James 1996: 132–133), agency (James 1996: 178–180, 185–186) and reality (James 1996: 159–160). In what sense, and to what extent, it is possible to establish meaningful connections between James's philosophic attitude and EM is itself potentially a controversial issue. The author's justification for proposing such connections stems from his own direct experience — in particular from the way in which James's discussion of 'pure experience' resonates with the issues involved in a detailed exposition of modelling with definitive scripts. In James's terms, the thrust of the exposition will be to make it plausible that the experience of EM 'knows' pure experience.

The paper is in two principal sections. Section 1 reviews Empirical Modelling (EM) principles and practice. Section 2 discusses William James's philosophic attitude of Radical Empiricism and the parallels that may be drawn between James's account of 'pure experience' and experience of EM. The paper concludes by identifying significant issues in knowledge management for which RE and EM may be seen as particularly relevant.

## 1. Empirical Modelling

Empirical Modelling (EM) is a term that has been introduced to describe principles and tools that support an unusual kind of computer-based model-building. The development of this approach has been the subject of an extended research programme at the University of Warwick that originated with the design of a notation for interactive graphics some 20 years ago. EM is unusual in that it represents a form of modelling oriented towards capturing 'state-as-experienced' and leads to the construction of computer-based artefacts that have no preconceived or formally circumscribed behaviour. It will be helpful to put this claim in context before presenting concrete evidence to illustrate it.

### 1.1 The nature of EM models

It is a commonplace fundamental notion of computer science that the significant semantics of a computer program is captured in the algorithmic behaviour that it implements. This notion leads on to the idea that all legitimate computer use is necessarily essentially concerned with specifying and implementing abstract behaviours. Whilst computer science acknowledges the need to design interfaces through which the user can direct or monitor computer behaviours, this is typically seen as beyond the remit of the core science of computing. Such a view of core computer science as fundamentally concerned with abstract mathematical concepts of computation and behaviour is curious in view of key trends in the historical development of computing practice.

The use of the computer — or more precisely of computer-related technology — to generate so-called virtual reality (VR) environments epitomises some of the most significant issues. If a VR environment is to have the qualities of a reality such as we experience in our everyday life, the character of the interface to the underlying computer model of state is a crucial rather than a peripheral issue. In a real environment, the user can observe the environment in ways that have not been preconceived, and conduct authentic measurements and experiments. It may be possible in principle to conceive how such a real environment can be implemented if we accept a reductionist view of reality and a logicist account of human intelligence. Such a conception is of little interest either to the computer practitioner or to the user, whose view of construction and use is guided by pragmatic concerns. For instance, we should not necessarily expect to have perfect knowledge in order to construct an environment to assist us in the task of 'knowledge management'. In any event, an environment in which only knowledge that has been previously encoded is recoverable is of limited use.

Viewed from the classical computational perspective, the idea of building computer-based models that offer the user more than has been consciously encoded by way of 'use cases' (cf. Jacobson et al. 1992) seems paradoxical. To see this from another perspective, it is helpful to consider other model-building activities, such as those associated with what Levi-Strauss characterises as *bricolage* (Levi-Strauss 1966). In bricolage, the modeller's concept of the artefact under construction develops in conjunction with the artefact itself: the modeller gains feedback from experience of the unformed artefact itself, and uses this feedback to guide its further development. To the extent that prototyping plays a part in computer program development, practical computer programming can be viewed in this light. In practice, feedback routinely affects the bricoleur's conception of the developing artefact in a much more radical way than prototyping affects software development. The bricoleur is not constrained by the pre-established conventions for interpretation that typically frame a software product: in contrast, the way in which the artefact is to be interpreted and used is in many aspects ill-defined and open to negotiation throughout its construction. EM may be seen as offering an approach to model-building with the computer that has these key characteristics in common with bricolage.

The classical view of computation is complementary to a classical view of knowledge. The concept of building programs that are optimised to serve a specific narrow function, and of encoding information in formal data structures, promotes a prosaic view of what knowledge and knowledge representation entail. Knowledge is seen as something to be possessed that can be expressed and recorded as a proposition, as in "I know the telephone number of staff member X". This view is appropriate in its proper context, as is implied by the use of the expressions 'perfect knowledge' and 'only knowledge that has been previously encoded' above. It makes sense to speak of recording the telephone numbers of all university staff in a directory, and — in the absence of any further context — it would be absurd to search this directory for those who are blue-eyed and can play the bassoon. On the other hand, it is evident that developing a VR environment involves much more than encoding abstract knowledge about a real-world domain. A key issue is how knowledge about the domain is reflected in the interactive experience that the environment offers to the user.

EM differs from classical computer-based model-building in its relation to both computation and knowledge. The distinction between an EM model and a computer program is analogous to the 'real-world' distinction between an open environment and a circumscribed procedure. To be familiar with an EM model is analogous to being familiar with a city; to be familiar with a computer program is analogous to knowing how to use the underground to get from one station to another. It is helpful to think of the EM model as something organic that grows and changes over time: there is no sense in which being familiar with a city is a limited notion — we explore and observe more of a city over time; the city changes; we change in our response to the city; we change the city. There is no reasonable sense in which familiarity with a city can be comprehensively preconceived; in any moment, we can at most testify to what is familiar in the particular aspect of the city of which we have immediate and direct experience. Though I may tell you with conviction that the cathedral is out of sight but lies just around the corner, I appeal to my memory and to my faith in the permanence of place, and know what only the act of taking us to the cathedral can confirm. An EM model is more general than a computer program in much the same way that 'knowing how to use the underground' sits within the broader framework of 'being familiar with the city'. The analogy also helps to illustrate the ontological distinction between the model and the program. If we take the permanence and reliability of a city environment and the generic and routine nature of our observation of underground transport for granted, then 'knowing how to use the underground' is a skill that can be viewed as independent of any particular city, that can be exercised without engaging with the total experience of 'becoming familiar with the city'. It is possible to imagine how a robot can be programmed with the stimulus-response patterns needed to use the underground, but much harder to conceive what might be meant by programming a robot to get to know a city.

The metaphor of "knowing the city" is helpful when understanding both the way in which an EM model is realised on a computer, and the philosophical stance it reflects on knowledge. As I look out of my study window at this moment, my view of the city of Coventry is limited to the roof of my neighbour's house and the sky above. In my imagination, I can trace the path from the house to the city, though the act of tracing this path is no part of my present direct experience. My knowledge of Coventry invokes the conjunction in my memory of all the direct experiences I have had over many years of different aspects of the city. There is no sense in which all these prior experiences can be taken as one experience, any more than I can be in more than one place at once. From this perspective, the 'perfect knowledge' on which we might aspire to base a classical computer-based model of the city is a mere chimera. What there is to be known of Coventry is more than I can ever experience, and my personal knowledge is established, maintained and revised dynamically through my ongoing interactions with it. Even my current limited view of the city is potentially open to such elaboration of knowledge, now that I notice the pigeon droppings on the crest of the roof, the silver sliver of a distant passing plane, and savour the taste of lukewarm lime juice cordial. As a source of direct experience, an EM model cannot compare with my everyday

Coventry environment in its richness, but it offers access to potential experience and knowledge with which I can engage in similar ways. In effect, it supplies an interactive artefact with which to trace a history of experiences that is similar in character to our everyday experience of purposefully and accidentally developing our knowledge of a city. In this activity, the computer serves two significant roles: it serves as a physical artefact, similar to the artefacts of bricolage, that generate direct experiences to which the modeller can make a creative response, and at the same time it allows the modeller to record and elaborate experiences as they are encountered in 'the stream of thought'. In these respects, the computer offers support for managing knowledge as characterised as 'interaction with memory'.

### 1.2 An illustrative case-study

Empirical Modelling principles and tools were not initially developed with the general characteristics of EM models, as described above, in mind. Our present understanding of the characteristics of EM models was arrived at after first appreciating the difficulties of providing a formal semantics for such models within the classical theoretical framework for computation. Two sources of inspiration have been particularly significant in reaching this understanding: a large body of practical work on model-building and tool development that has been carried out largely by computer science students at Warwick over the last 15 years, and the philosophical writings of William James. This subsection will give a brief sketch of practical EM with reference to a model of a lift initially developed in a summer vacation project in 1994. (The lift model and the EDEN interpreter which is needed to run the model can be downloaded from the EM website). The next section will then discuss EM in relation to William James's philosophic attitude of Radical Empiricism (James 1996).

Figure 1 depicts our EM model of the lift as it might be displayed in one particular state. The lift car appears on the left, and the five floors that the lift visits on the right. The stick figures are used to indicate the current locations of lift users: a label attached to the head of each stick figure identifies the list of users at each location. The bold vertical lines between the lift shaft and the floors represent the status of the doors between the lift car and each floor: the door on the fourth floor is currently open. The small boxes to the right of the floors and to the left of the lift car are call buttons, and those that are currently selected are indicated in bold.

To interpret Figure 1 as associated with a direct experience of an actual lift, it is helpful to think of looking at a glass lift at the side of a five storey building. Whether this is representative of the direct experience that is of primary interest to the modeller depends upon the role in which the modeller interacts. Examples of possible roles that the modeller might play, possibly concurrently, include: a lift user, designer or analyst, or perhaps even a story teller for whom a lift is a significant location (e.g., as the venue for murder in a detective story). In practice, if we have any familiarity with lifts (even if we are in the position of user X in Figure 1, with limited access to the actual current status of the lift), we bring to our interaction with them a general concept that there are floors and users and a lift car that moves vertically between them, and may be expected to visit the ground floor at some stage in response to pressing the call button.

In our modelling tool (`EDEN`), the current status of the lift is determined by the current values of its characteristic observables. The names of some of the key observables, as they are recorded in our model-building interpreter, are:

| | |
|---|---|
| `_liftfloor` | – where is the lift? [on which floor: 1–5] |
| `_open3` | – is the door open at level 3? |
| `_car2` | – is button 2 in the car selected? |
| `_button4` | – is the call button on floor 4 selected? |
| `locX` | – where is user X? [0: in lift, 1–5: on which floor, 7: nowhere] |
| `_inliftX` | – is user X in the lift? |
| `_destX` | – where is user X intending to go? |

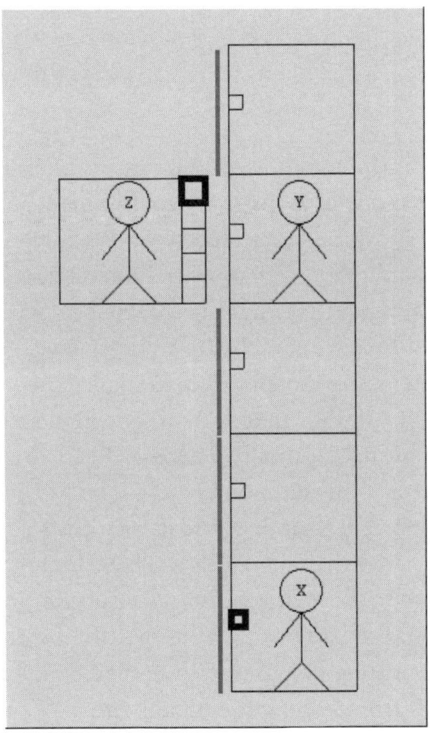


**FIGURE 1.** EM model of a lift

The values of most of these observables are directly reflected in features of the display. For instance, `_liftfloor` determines the position of the lift, `_open3` determines how the line that represents the door on floor 3 is depicted and `_button4` determines whether the call button on floor 4 is highlighted. It is characteristic of modelling with `EDEN` that these features of the display can themselves be construed as observables whose values are specified by definitions resembling the definitions that relate the cells of a spreadsheet. Taken together, the definitions of observables, whether explicitly by values, or implicitly via formulae, form a 'definitive script' that specifies the current state of the model. The term 'definitive notation' is used to refer to the underlying notations used to formulate a script. `EDEN` includes a definitive notation for defining the screen layout and for defining planar line drawings. The use of these notations is illustrated in Figure 1.

The definitions in a script express expectations about how the values of observables are linked where 'atomic' changes are concerned. For instance, when the lift moves, anyone in the lift moves with it as part of one and the same state change. The relations linking these changes are called 'dependencies' between observables. For example, using `carpos` and `posX` respectively to represent the positions of the lift car and of user X (as depicted by 2D coordinates in Figure 1), definitions of the following general type express dependencies:

```
carpos is {liftshaftL, (_liftfloor-1) * floorheight}
posX is if _inliftX then carpos else floorX
_inliftX is (locX == 0)
```

The use of a definitive script to represent state-as-experienced is the fundamental technique by which a modeller constructs environments for interaction in EM. Such an environment is typically oriented towards expressing the perspective of a particular agent where both observation and interaction are concerned. In this context, an agent is anything that might be deemed to be responsible for changing the current state of the EM model. The archetypal agent is the modeller who acts to change the state of the model through manually entering redefinitions of observables into the `EDEN` input window. As will be discussed and illustrated later, the lift users and the lift itself can also be viewed and in various different ways animated as agents with some autonomous capacity to change state. In viewing the EM model, it is important to understand its primary role as a representation of state-as-experienced, rather than merely as a vehicle for automating conventional lift behaviour. To return to the analogy introduced above, animated behaviours in EM are in the first instance like bus tours of the city: by default, on such a tour, we observe the city according to a pre-programmed plan, but there is nothing in principle to prevent us from leaving the organised party and continuing our exploration independently.

The idea that the lift model captures state-as-experienced can only be appreciated by considering the possible modes of interaction with it that are open to the modeller. To understand this, it is helpful to reflect on the distinction between the abstract interaction that is involved in using the underground and that involved in making the same journey above ground. In the former context, the traveller can get by by pattern matching, relying on the way in which the underground environment itself is built to encode knowledge about locations and directions in place names, keywords and icons. In the latter context, the personal knowledge of the traveller — their ability to recognise landmarks, to maintain a sense of direction, to devise strategies for crossing roads, to be able to consult other people — is paramount: though the environment itself has embedded knowledge in the form of signposts and street-names, the logistics of their use are by no means fully systematised. In practice, both opportunistic and exploratory human interpretation and pre-coded knowledge play their part in navigation in a city environment. In a similar way, interaction with the EM model of the lift can either give prominence to how human intelligence informs state transition or support routine and automatic interaction. What is more, commitment to changing the model is not necessarily required in playing these different interactive roles, and either can be highlighted according to purpose.

The openness of the modeller's agency in EM is the platform for intelligent interaction. The modeller is free to redefine any of the observables in the script subject only to being able to interpret the consequences of such redefinition. As a simple example, a redefinition such as

```
carpos is {liftshaftL, 0}
```

relocates the lift car at the base of the lift shaft as if it were out of service. Note that this redefinition in some sense 'deals intelligently with user Z': in the context of Figure 1, it would move user Z within the lift no matter where the lift was placed. Actions of this open-ended nature are analogous to the unconstrained actions that the traveller can make in the city environment, which may result in reaching the destination inadvertently, might lead to getting lost or to being run over.

To illustrate how knowledge about meaningful interaction can be exposed without being encoded in an EM model, the lift model has been adapted and used to animate the unusual instance of real-life lift use described in Box 1. For this purpose, the modeller has only to compile a sequence of elementary redefinitions of observables to reflect the actions of the users and the lift as they occur. By playing through this sequence of redefinitions, the modeller can visit each situation as it arises, and reflect upon the perceptions and motivations of the various agents involved. The modeller can then construct a narrative to document their informal understanding of the scenario. The `EDEN` interpreter provides a procedural

I am at a conference in the Netherlands.
I arrive late at night and hardly notice where my room is.
Next morning, I notice that my room is on the top floor.
I walk down to breakfast thinking about my talk later on.

After breakfast I meet two other delegates X and Y.
We get in the lift to return to our rooms.
X presses the button for floor 4.
Y says he is on the floor above X, and selects floor 5.
Since the top button is selected, I don't press a button.

We talk as we ascend. The lift stops. The door opens.
The floor numbers aren't clearly marked.
I say to X 'this must be floor 4' — he gets out.
Y and I carry on talking.

When the lift next stops, the floor is still unclear.
I say to Y 'X is on the floor below you; this is your floor'.
Y gets out. I think something is not quite right.
I think 'is this the top floor?' and 'should I get out?'.
I'm unsure, but notice that the button for floor 5 is still lit.
I proceed to the top floor which is the next floor, floor 5.
When I get out of the lift, I can't find my room.

There's no room where my room is on floor 5.
I walk down to floor 4, and pass Y on his way to floor 5.
When I reach floor 4, I meet X coming up from floor 3

*How did I manage to get all three of us to the wrong floor?*

**Box 1.** Travails in a lift

construct for playing out this sequence of redefinitions as if they were being entered one-by-one by the modeller. Such a sequence can be accompanied by a commentary that gives more insight into the complex combination of observation and logic that guides each lift user's actions.

By way of context, the events described in this scenario occurred some fifteen years ago, and have been a source of puzzlement to the author ever since. Constructing the model has finally enabled me to reconcile my assured personal knowledge of the circumstances of the interaction with a plausible objective account of the behaviour of three sane lift users. To dramatise the distinction between these two perspectives on the events, I shall complement the personal account in Box 1 with a further analysis in which my role is described in the third person, as that of 'user I'.

Figure 2 depicts a critical moment in the scenario, at which it becomes clear to user I that there is some conflict between what I observes and the way in which

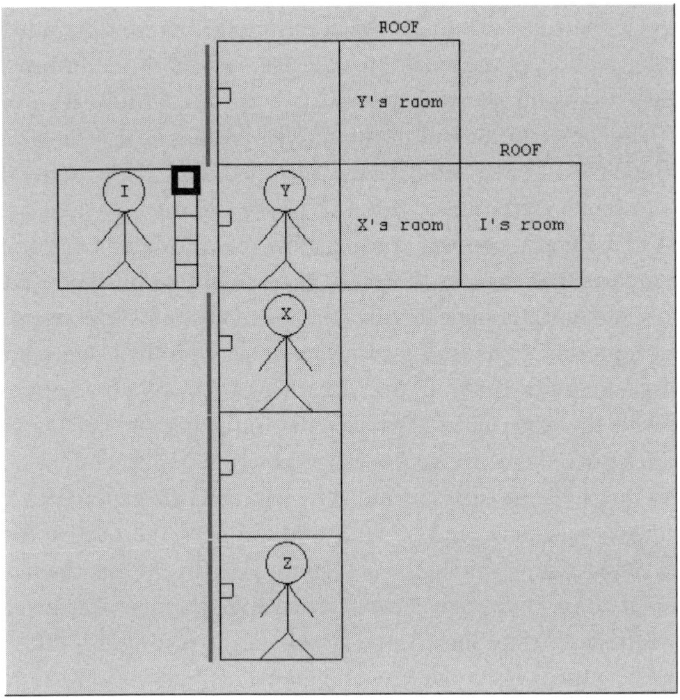


FIGURE 2. Embellished EM model of a lift

I had conceived X and Y's interaction with the lift up to this point. I supposes that his room is on the top floor, and realises that the lift has not yet reached the top floor. On the other hand, on entering the lift, I observed the selection of just two buttons by co-users X and Y, and noted that one of these corresponded to the top floor. It would seem from this that I should have got out at the same time as Y. It was just after Y got out of the lift that I realised that perhaps he had misled X, and that the lift had in fact stopped at floor 3 when called by an impatient user Z. At that point, it was apparent to I that both X and Y had got out at the wrong floors, but not that he himself was also about to travel upwards to the wrong floor. The extra detail of the profile of the building that has been introduced in Figure 2 represents the other element in the situation by which I himself was misled into supposing that his room was on floor 5 rather than floor 4: in the vicinity of I's room, floor 4 was the top floor. The pragmatic nature of the decision that I makes at this point, so different in character from the pure reasoning that is sometimes imputed to human agents (cf. the Mensa problem discussed in Beynon 1999), is emphasised by the real time constraints on decision imposed by the lift itself.

Figure 2 can be interpreted as the particular specialisation of the vanilla model of the lift depicted in Figure 1 that I ideally should have had in mind in using the lift. In practice, I was not able to construct this model in real-time, and acted with a different conception of the situation. A crucial aspect of the EM model is that it relies essentially upon physical artefacts to account faithfully for I's misunderstanding. The geometric content of the model, limited as it is as far as realistic details of the physical environment are concerned, has rich experiential significance. It is to remembered observation of physical models such as are depicted in Figures 1 and 2 that we appeal in using a lift, and the experience of interacting with these models that exposes their role in representing our knowledge of lifts. Needless to say, many attempts have been made to abstract the essence from such experiential representations in logicist approaches to knowledge representation (cf. Ginsberg and Smith 1988). In practice, there are many other pertinent issues connected with the experience of lift use that influence the behaviour of users, many of which are only tacitly acknowledged even in our account of the scenario described in Box 1. These issues include: the nature of the experience — the time it takes a lift to move between one floor and the next, the relative time it takes to ascend 3 rather than 4 floors, the acceleration of the lift car; the nature of the auxiliary observation that might have affected user responses, as determined by the status of the buttons of the lift, the visual cues to distinguish different floors, whether and how the users' room keys and the lift car buttons were numbered; the implicit conventions framing the lift design and use, such as whether the ground floor was deemed to be floor 0 or floor 1, the fact that the buttons in the lift were ordered top to bottom in a 1-dimensional array in correspondence with the five floors, or that a call button in the lift was visibly de-selected when the door opened at the corresponding floor.

As the above discussion illustrates, the EM model of the lift offers insight into real world state-transitions about which the automation of the perfect lift user can give no information. Interaction with Figure 2 offers a plausible explanation for corporate behaviour that is absurdly different from the ideal outcome to be expected, but it also shows that — taking appropriate observational and experiential factors into account — the individual actions of the users are not so very far from sensible as they might appear. Logic alone cannot account for the scenarios that lie near to normality 'in the neighbourhood of sense'. In exploring such scenarios fully, there is no alternative but to allow the modeller to engage as freely as possible in the state-changing activity. In the spirit of Gooding's definition of the term, an EM model can be interpreted as a *construal* (Gooding 1990). In constructing such a model using EM principles, our aim is to embody the patterns of observables, dependency and agency that we deem to be characteristic of a situation. The product of this activity typically has a personal, provisional and particular character that is intimately connected with the ways in which we choose to interact with it. In all these respects, it is unlike a formal specification.

In developing a construal, the modeller is not only concerned with modelling state change as it is observed from an external perspective. A further analysis of the sequence of redefinitions that are made in realising the scenario in Box 1 shows how certain groups of redefinitions can be attributed to different agents. The movement of the lift and the manipulation of the lift doors are part of the automatic behaviour of the lift system itself. Entering and leaving the lift and the selection of call buttons inside and outside the lift are actions for which the lift users are responsible. To refine the model so that it better reflects our understanding of normal lift operation, it is appropriate to distinguish between realistic transitions and transitions that are pure fantasy, such as might involve simulating a user entering the lift before the door is open, or the lift jumping between floors. EM includes techniques for enriching the lift model so as to give special prominence to construals suited to different perspectives and purposes, some of which may include automated behaviour, or reflect several viewpoints concurrently. It is important to realise that the specialisation of an EM model does not involve excluding singular actions on the part of the modeller. In some circumstances, such as when the lift cable breaks, it may be appropriate for the lift to exhibit abnormal behaviours that can only be simulated by the intervention of the modeller in the role of a super-agent. The `EDEN` interpreter is designed to support the opportunistic interaction by the modeller that is required for this purpose, but whether such privileges for interaction are exercised is entirely at the modeller's discretion. A brief review of some relevant EM techniques is given here — for more details, consult the references on the EM website.

In elaborating construals, a significant role is played by an informal special-purpose notation, called `LSD`, that can be used to classify the observables in a situation with respect to an agent. The aim of such a classification is to document the roles that observables play in determining the interaction of an agent. Listing 1 is an `LSD` account of a person X in the role of a prototypical user of the lift depicted in Figure 2.

In this account, the observables `loc[X]` and `dest[X]` refer to the location and destination floor of user X respectively. The values of these observables are defined according to the same conventions used to define `locX` and `destX` in the scripts discussed above. The location of X is classified as a *state* for X as a person, since observation of this location is meaningless in the absence of X, and the destination floor of X is classified as a state for X in the role of lift user, since the concept of X's destination floor is meaningful only when X acts in this role. Whether or not X is currently playing the `liftuser` role is reflected in the boolean value of the special

```
agent person (X) {
   state loc[X]
   role liftuser {
      state dest[X]
      oracle open[*], dest[X], loc[X], pos[X]
      handle loc[X], dest[X]
      derivate
         pos[X] is if loc[X]==0 then liftfloor else loc[X],
         LIVE_liftuser[X] is 0<=loc[X]<=5
      protocol
         loc[X]==0 and open[dest[X]] and pos[X] == dest[X]
               -> loc[X] = pos[X],
         true  -> dest[X] = i,          (1<=i<=5)
         ...
   }
   ...
}
```

Listing 1. An LSD account of a prototypical lift user

observable $\mathtt{LIVE}_{\mathtt{liftuser[X]}}$, which is true or false according to the current value of loc[X]. As a lift user, X has access to certain observables as *oracles*, either in the sense that she can observe them directly, or that she can observe them at certain times, and may retain some notion of their current values. She typically knows her destination floor. She also knows whether she is in the lift, or at a floor, as determined by the oracle loc[X]. If X is in the lift, she has some notion of what floor the lift is currently at, as represented by the observable pos[X]. The observable pos[X] is classified as a *derivate* because of the dependency that defines it in terms of the current lift position and the value of loc[X]. The observables over which X can conditionally exercise control are classified as *handles*: they include X's location and destination. The conditions under which X can redefine these observables are set out in her *protocol*. The two example privileges specified for X indicate that if X is in the lift, the lift is at her destination floor and the door is open, then X can step out of the lift, and that moreover X can change her mind about her destination floor at any time.

An LSD account is not in general to be interpreted as a specification, but rather as documenting the characteristics of observables as they are experienced by the modeller in the real situation, and (if the EM model is sufficiently convincing) as they are experienced in the associated EM model. In effect, an LSD account is intended to complement an artefact with which the modeller can interact, and is not to be viewed as an alternative form of representation. An LSD account can play a significant role in guiding the development of several different kinds of EM model for a multi-agent system such as a lift. These include:

**Concurrent modelling of user perspectives on lift use.** An LSD account can be seen as describing the system as viewed from the perspectives of its various users. These views can then supply the basis for a distributed EM model in which the personal construal of each agent is modelled as a client in a client-server configuration, and the corporate behaviour of the agents is developed by managing their interaction via the server. For instance, using a distributed variant of the EDEN interpreter, it would be possible to set up an EM model similar to that depicted in Figure 2 on a server, and to create EM models on four clients to recreate the interaction of the lift users X, Y, Z and I in the scenario as it appears from their four different perspectives. In such a model, there will be nothing that corresponds to the lift users X, Y and I from the perspective of user Z, since Z merely presses the call button on floor 3 then descends to floor 2. The models for X, Y and I will be based on different variants of the prototypical lift user account given above. For instance, the oracles `dest[X]`, `dest[Y]` and `dest[I]` have to be construed quite differently: user X has a correct perception that `dest[X]` is 4, but (when encouraged by I), supposes that `pos[X]` is 4 when it is in fact 3; user Y has an oracle to `dest[X]` which takes its value from the floor on which X gets out, and interprets `dest[Y]` as a derivate defined by `dest[X]+1`; user I has the number of the top floor of the building as an additional observable topfloor and interprets the oracle `dest[I]` as a derivate defined by `topfloor`. A distributed model of this nature can provide a more vivid reconstruction of a scenario than the basic EM model depicted in Figure 2, as has been illustrated elsewhere in our previous research in the reconstruction of historic railway accident developed by Sun (Beynon and Sun 1999).

**Modelling the lift as a reactive system.** An LSD account can be used in analysing the stimulus-response behaviour of automatic agents as they are construed to respond and act upon observables in their environment. Activity of this nature has a fundamental role in experimental science. It is also an important aspect of reactive systems development, where it is associated with the exploration and specification of context that precedes the design of control software. In applying EM principles to such activities, the most significant feature is making the connection between model-building from a personal subjective perspective and what is interpreted as objective observation of a system by an external observer. A full discussion of the technical issues involved is beyond the scope of this paper — it has been one of the primary concerns of the EM project as a whole (cf. Beynon 1999; Beynon and Sun 1999; Beynon, Rungrattanaubol, and Sinclair 2000; Rungrattanaubol 2002). The basic concept is that of treating the activity of automatic agents as it were being carried out by a human agent with the appropriate perceptions and state-changing capabilities. By way of illustration, the lift system can be viewed as automatically carrying out the sequences of redefinitions required to reset the call buttons and open and close the doors on visiting a floor to which it is called. The stimulus for this operation is the presence of the lift car at a floor that is associated with a selected call button. An `LSD` account of this role of the lift system would (for instance) identify oracles — such as the status of call buttons, handles — such as the status of the doors, and include an action to manipulate the doors and call buttons appropriately in its protocol. A similar analysis can be used to develop a protocol to prescribe the motion of the lift. The `EDEN` interpreter includes features, such as procedures that are triggered by changes to specified observables, which can be introduced into the EM model of the lift to automate the lift system protocols. Such features make it possible to implement a lift simulation through an incremental process of extending the EM model in which the modeller's role involves shaping the behaviour to accord with realistic observation and interaction. The delay after which the doors close can be adjusted, for instance, and the selection of buttons by lift users simulated by direct mouse actions. The significant point here is that the `LSD` account has no formal operational semantics, but documents actual interaction with a computer-based artefact for which behaviours can be developed in much the same incremental and empirical fashion that an engineer might construct a prototype. In this process of empirical refinement of the EM model, the scope for extension is open: simple extensions that feature more realistic lift motion, implement a lift scheduling algorithm, and introduce prototypical users based on the `LSD` account in Listing 1 can be found in the `liftBeynon2003` directory of the EM archive. Further extensions for this model might involve introducing greater physical realism by way of modelling lift car velocity and acceleration, adding 3-dimensional visualisation, or linking the model to special-purpose hardware that could simulate the impact on the user of forces generated by the lift motion.

The above description and illustration of EM sets the scene for the discussion of Radical Empiricism that follows. To fully appreciate this discussion, it is most helpful to have some experience of the nature of EM as a practical activity. Without such experience, it is difficult to appreciate the conceptual distinction between traditional computer programs and EM models that is described in Section 1.1. It is in particular important to realise that all the lift models discussed in this section are to be regarded as part of a single open-ended conceptual process of exploration that can be seen as resembling the exploration of a city. Each model is informally associated with a particular way of viewing a lift situation and of organising the transitions between one situation and the next, but each is apprehended — even when, left to itself, it is executing a particular pattern of behaviour — state-by-state, in such a way that the modeller can choose to intervene to redirect the experience perhaps with a view to its reinterpretation. In this respect, the individuality of the different lift models is not defined objectively, but only with reference to what experiences are coherent for the individual modeller.

## 2. Radical Empiricism

William James's *Essays in Radical Empiricism* was first published in 1912. The potential relevance of James's 'philosophic attitude' to a discussion of alternatives to a logicist framework for knowledge representation is apparent throughout these essays. In 'The World of Pure Experience', for instance, when discussing the philosophic atmosphere of his time, James refers to "a feeling that [the extant school-solutions] are too abstract and academic", and goes on to write: "Life is confused and superabundant, and what the younger generation appears to crave is more of the temperament of life in its philosophy, even though it were at some cost of logical rigor and of formal purity" (1996: 39). Empirical Modelling is motivated by a perceived need to develop methods of computer-based modelling that can do more justice to life in its confusion and superabundance than can the rational formal accounts of agent interaction that underlie typical computer programs. The distinction between these different views of human agency has been illustrated above when contrasting the farcical scenario of lift use associated with Box 1 with the prosaic and predictable behaviour that is attributed to prototypical lift users in Listing 1. This section explores other respects in which the principles and techniques of EM can be related to thinking developed by James (1996). Our overall aim is to make it plausible that James's philosophical stance supplies a foundational framework in which to examine issues that seem paradoxical from more conventional philosophical perspectives.

### 2.1 Philosophical foundations for EM

Some background motivation for our discussion can be found by thinking in general terms about what kind of philosophical foundations are appropriate for EM. One of the most characteristic activities in EM is the construction of a computer-based artefact (for instance, the lift model in Figure 2) that embodies patterns of observables, dependencies and agency that can be identified in an external situation to which the artefact refers (for instance, the specific use of the actual lift described in the scenario in Box 1). From a traditional computer science viewpoint, where there is an underlying presumption that all computer-based modelling can be accounted for by using the universal abstractions that rest ultimately upon the classical theory of computation, it is usual to propose that EM reduces to classical programming through a correspondence of the following general kind: an observable is a procedural variable, a dependency is a constraint relation, an agent is a sophisticated abstraction such as an active object. The renunciation of each of these proposed reductions has been the focus of special attention in our previous work (see for instance the discussion of variables in Beynon and Russ 1991, of constraints in Rungrattanaubol 2002 and of agency in Beynon 1999). Broadly, our counter to this suggestion is similar in all three cases: that the notion of 'observable' refers to a feature of a situation that is experienced as having an identity and current status or value; that a 'dependency' is more than a perceived abstract relationship between values of variables and expresses the modeller's expectations about the immediate consequences of changing the values of observables in a situation; that 'agency' entails a potential for action that is of a truly experimental character, in that the possibility of taking the action has not been preconceived, and that no prior commitment to the possible interpretation of its consequences has been made. What links each of these counterproposals is the context in which our interaction with the computer model is conceived to be occurring: a context resembling that through which I as-of-now experience the city of Coventry through the sight of my neighbour's house and the sound of his lawnmower.

Our rejection of a conventional interpretation for EM has another significant element that is closely linked to Brian Cantwell Smith's critique of classical computational semantics (Cantwell-Smith 2002). The proposed reduction of an EM model to a classical program purports to attribute an abstract behaviour to the EM model, in accordance with the ways in which variables, constraints and active objects might be used to specify behaviours in a conventional approach to model-building. The development of an EM model does not rely upon the identification of an abstract behaviour that can be embedded into the environment through formal symbolic associations. This is to revisit our previous observation, that an EM model does not in the first instance specify an activity, like travelling about a city using the underground, for which — thanks to the traveller's training in symbol recognition, and the careful engineering and signposting of the underground environment — limited experience of the city is required. On the contrary, the EM model offers itself to the modeller as a state to be experienced, where the correspondence between the features of the model and those of its referent are to be directly established, explored and enhanced through interaction. This perspective has to be understood with reference to a philosophical position that assumes no given absolute knowledge, in the same spirit that (in my current context) I cannot be absolutely sure that all I remember of the city of Coventry will be there to experience when I set off to visit the railway station.

As Cantwell Smith (2002) observes, the inadequacy of the classical view of computation is conspicuously exposed in emerging computing practice. The wide range of applications for EM principles that we have identified to date highlights both the potential of EM and the challenge of understanding its semantics. The pragmatic view of the semantics of modelling 'real-world objects' that serves well enough in traditional engineering design can to some extent be adapted to an exercise such as using EM to model an actual lift. The concept of a direct correspondence between experience of an EM model and the experience of the real-world situation it represents requires more justification when the EM model is a spreadsheet, and the situation a financial scenario. It is also more difficult to argue for a direct correspondence between observables for an EM model to represent a lift that is under design and has yet to be built. Further complications arise if we consider the status of the EM model of a lift that we might create in a virtual reality, where the relationships between observables are no longer subject to familiar physical laws and constraints. EM has been used to build models that have the experiential characteristics of the data structures (such as the heap) that lie behind standard algorithms (such as heapsort), and thereby to generate an environment in which to explore the design of algorithms (Beynon, Rungrattanaubol, and Sinclair 2000). In other contexts, it is apparent that the use of EM principles is not directly concerned with issues of external representation. For instance, our tools now enable us to design new definitive notations — in particular, for graphics — within the same paradigm that we use to construct models, and — for such a notation — successful design is concerned with how the syntax of the new notation affects the correspondence between the structure of a script and the graphical image that it produces on the computer screen.

A feature common to all these applications of EM is the construction of a computer-based artefact that in some respects has the qualities of an instrument (cf. Beynon et al. 2001). The term 'instrument' is used here to express the idea that there is a reliable correlation between the interactions of the modeller with artefact and its associated changes of state. This correlation may be explicitly engineered by the modeller, or learned through skill acquired in mastery of the instrument. Explicit engineering is prominent in the case of a financial instrument such as a spreadsheet, where the definitions of cells are contrived to express known relationships between observables. Skilful interaction is more prominent in relation to a musical instrument, such as the violin, where the tiniest nuances in the movement of the bow can be used to control the sound generated. In broad terms, the account of EM that we have sought to develop in our project represents EM models as instruments simultaneously under development and in use by the modeller, in which both explicit engineering and skilful interaction have a role to play. Through explicit formulation of dependencies and through experimental redefinition, the modeller develops an understanding of how interactions with an EM model reliably effect changes to its state. In this account of EM, the interaction with the instrument does not acquire meaning as a result of a complex abstract process of off-line decoding; it directly evokes a parallel experience because of the perceived similarity between the effects of interaction with the instrument and the familiar effects of interaction in another context. It is in just this manner that the movement of geometric elements of the simple drawing in Figure 2 evokes familiar interactions with an actual lift.

The correspondence between one experience and another that underpins EM is different in character from the realistic modelling of behaviour and appearance that is commonplace in routine prototyping of software and engineering systems. For instance, our primary concern is not with simulating the lift dynamics as accurately as possible by analysing the forces acting on the lift in detail and applying Newton's laws, nor with developing a virtual reality model that imitates the user's visual and sensory experience as faithfully as possible. As the lift model illustrates, EM can serve to establish such similarities, but the connection between an EM model and the situation to which it refers is more direct and primitive in nature. The correspondence between interactions with the EM model and interactions with its referent is itself a matter of immediate experience, subject to confirmation, exploration or possibly even refutation by the modeller 'as of now'. A correspondence of this nature is rooted in the notions of 'observable', 'dependency' and 'agency' as they have been discussed above: it has a concrete and dynamic rather than abstract and static quality. In immediate experience, the agency of the modeller has the capacity to confirm or confound expectations, to create or destroy observables, to make or break dependencies. In these respects, it resembles the agency of the experimental scientist, whose actions can be used to test her construal.

The potential role that language and symbolic conventions can play in this context is a delicate issue. It is self-evident that our apprehension of correspondences in immediate experience can be mediated by language, as when we respond to the value in the column of the spreadsheet that is headed 'balance' or 'profit'. It is quite another matter to argue — as some philosophical traditions appear to do — that language plays an essential role in all correspondences between experiences. A core idea in EM is that not all correspondences between one experience and another can be established by symbolic conventions — at some level, correspondences must be made through direct experience without reference to language. The expectations of the experimental scientist may well entail much sophisticated theory, but the primitive correspondences that provide the grounding for such theory are arguably beyond words and equations. Experience in applying EM principles and tools suggests the further hypothesis that agency, dependency and observation have a fundamental role in such primitive correspondences.

## 2.2 Radical Empiricism from an EM perspective

Experience of EM, and reflection about its semantics, endorses a philosophical position that has strong points of connection with William James's Radical Empiricism. The primitive correspondences between experiences at the core of EM are necessarily personal experiences, and in the first instance are associated with subjective and provisional knowledge. This establishes priorities that are in line with those of RE: to account for logic and theory in terms of observation and experiment, rather than to account for observation and experiment in terms of logic and theory. Practical experience of EM reinforces this perspective, demonstrating how EM can be accompanied by transitions from personal to public, subjective to objective, and provisional to assured (cf. Beynon 1999 for more discussion of these issues). This section discusses how some key ideas of RE are helpful in elaborating on what is involved in EM.

James (1996:42) remarks that, in RE, "the relations that connect experiences must themselves be experienced relations, and any kind of relation experienced must be accounted as 'real' as anything else in the system". One of his primary concerns is that traditional empiricists emphasise the disjunction in experience "leaving things permanently disjoined", whilst the rationalists remedy this disjunction "by their Absolutes …, or whatever other fictitious agencies of union they may have employed" (p. 51). For James, both conjunctive and disjunctive relations should be deemed to be given in experience. Amongst these, he includes "the most intimate of conjunctive relations, the passing of one experience into another when they belong to the same self" (p. 50).

James's outlook helps to explain the difficulty we have encountered in understanding and communicating the nature of EM models. The typical computer scientist has a natural desire to attribute a characteristic set of discrete states and behaviours to the EM model of the lift in Figure 1, and to view it as a structure or system. In the mind of the modeller, the character of the EM model is more elusive. We can experience the state of the lift model as it is depicted in Figure 1, but interpret this in relation to all the other experiences of the states that have brought the model to this point. By the same token, what we experience in the current state of the model stands in an open, yet to be explored, relation to other experiences we can get by interacting with the model. What we understand by the model is an unfolding conjunction of experiences that cannot be circumscribed but is apprehended as a single relation.

An important aspect of James's analysis is the implicit emphasis it places on the authenticity of the personal experience of the observer. This distances RE from the perspective with which empiricism is ordinarily associated, where the 'reality of the given world' is seen as the primary source of knowledge. As our practical experience of EM has shown, it is implausible that we can give a good account of EM without regarding reality as constructed by experience. Though it is convenient in describing EM to adopt everyday terminology, and speak of 'the modeller's state of mind', and 'the real-world situation', this should not be understood as subscribing to a Cartesian dualism. This accords with James's (1996: 141) observation that "subjectivity and objectivity are affairs not of what an experience is aboriginally made of, but of its classification. Classifications depend on our temporary purposes ...". In observing the development of the EM model of the lift, we are led to think quite as much about the evolving state of mind of the modeller as about realistic changes to the state of the actual lift, whether these take the form of a movement from one floor to the next, or adding labels to the call buttons in the lift car. This is keeping with our perception in EM that, contrary to the received view that a computer model of lift should only — perhaps even can only — be constructed with a specific goal and behaviour in mind, our EM model of the lift is somewhat neutral to purpose. Certainly, the experiences of the lift model that would be generated through the interaction of modellers acting in the roles of lift users, lift designers, lift analysts, or story tellers would be quite different, and reflect many different kinds of conjunctive relation in their minds.

Of particular relevance to EM is James's contention that "the first great pitfall from which [RE] will save us is an artificial conception of the relations between knower and known" (James 1996: 52). This issue relates directly to the discussion above concerning how correspondences between experiences are established. The artificiality to which James alludes here stems from what is perceived by other philosophers as the 'indefeasibly dualistic' structure of experience, as James puts it. For James, there is no such duality: "All the while, in the very bosom of the finite experience, every conjunction required to make the relation intelligible is given in full" (James 1996: 53).

James's philosophical position is helpful in understanding the nature of EM. As has been argued above, the fact that the EM model of a lift stands in a special relation to an actual lift (viz. in what James (1996: 52) characterises as "the relation between knower and known") is not illuminated by classifying the one as a 'mental model' and the other as a 'real-world object'. What matters is that they are two portions of experience that are related in a way that is itself experienced by the modeller. This leads us to a characterisation of the modeller that matches our experience of EM activity well: that of an agent who generates an experience that knows another. In the same spirit, EM principles and tools can be seen as assisting the generation of this experience. In this context, our choice of the word 'model' belies our faithfulness to James's account of knowing, as it is commonly associated with what James rejects by way of "representative theories" (1996: 52).

The simplicity of the relation of knowing, as James describes it, is consistent with our experience of how readily EM models can be combined and reinterpreted. James rejects the duplicity of experience that distinguishes 'consciousness' from 'content':

> Experience … has no such inner duplicity; and the separation of it into consciousness and content comes, not by way of subtraction, but by way of addition — the addition, to a given concrete piece of it, of other sets of experiences, in connection with which severally its use or function may be of different kinds (James 1996: 9).

This explanation accounts for the way in which the object of study in one exercise in EM seamlessly becomes a part of the EM model in another; for the fact that an EM artefact, like an artefact in bricolage, can be developed without a referent in mind; for the ambiguity about what features of an EM model serve a significant representational role (cf. the bold lines indicating the lift doors, the size of the call button panel in the lift, the presence of the roofs in Figure 2). In each of these contexts, the precise character of the EM model is only shaped by the nature of the interaction of the modeller as it unfolds at her discretion — in particular, by how the values of observables are interpreted and changed, and how these changes to observables are interpreted. By way of illustration, a student who used EM to simulate bread baking in an oven chose to represent the oven by borrowing a simple line drawing to represent the floor plan of a room (see `roomviewerYung1991` from the EM archive), and labelled his simulation by substituting 'Bread baking simulation' for a warning message that notified the user when a table obstructed the door. This meant that the identifying label could be changed by relocating an invisible table — a functionality that the student did not document, and was unacknowledged in his personal interpretation of the model.

Though both RE and EM share a central concern for rooting knowledge in personal experience, their agendas are quite different, and explicit connections are hard to make. Despite this, RE is helpful in developing a deeper understanding of the primitive concepts of EM, for which formal logical foundations cannot be supplied.

James's account of pure experience provides a most appropriate setting in which to explain the notion of an observable in EM. The appropriate sense of being an observable in EM is 'having an identity' and 'having a value that can be directly experienced', where the term 'is directly experienced' refers to the capability of the human interpreter in the given context to apprehend immediately. There is a most significant distinction between this notion of an observable and what traditional empiricism might deem to be an observable. To the infant, the symbol '2' on the lift button is a mere pattern of sensation that signifies nothing; to a young child, it will be associated with the idea of 'some pair of objects'; to the competent lift user, it is known to refer to a specific floor. James's account of knowing dispenses with the idea that "seeing the symbol '2'", "thinking of two floors" and "imagining going to the second floor" are categorically different kinds of experience whose association in the mind of the lift user must be explained by "fictitious agencies of union". When situated in the lift, the lift user experiences the conjunctive relation that connects all three elements of his experience of the button labelled '2'. The uniform way in which observables are treated in an EM model, regardless of the level of sophistication of the observation involved so long only as it is immediate (cf. labelling the call button by 'II', or "The smallest prime number"), is consistent with James's outlook. The fact that in EM, as in life, the scope and significance of such observables is dynamically established as experience is acquired — perhaps from moment to moment, as a child might be taught *in situ* to connect the symbol '2' with a pair of floors — is also in keeping with James's conception of knowledge: "Why insist that knowing is a static relation out of time when it practically seems so much a function of our active life?" (James 1996: 75).

In EM, there is an intimate relationship between the notions of dependency and agency. Dependency is in effect a way of binding together all the consequences that are deemed to be an integral part of a single action. In EM models, this concept is realised practically through modelling with definitive scripts (cf. Rungrattanaubol 2002), where a typical atomic action involves redefining an observable or introducing / removing a cluster of observables. An analysis of the notion of dependency reveals a number of respects in which it is related to agency. Whether a dependency is identified in a particular context in general depends on the perspective and agency that the modeller has in mind. The position of the lift car might be seen as dependent on the position of the lift cable, but this view is entirely appropriate only if we assume an idealised mechanical model (e.g., making no allowance for the initial extension of the cable under load), discount the discrepancies between the exact positions of the lift car within the tolerances allowed in the design of the lift shaft, and decide not to model the behaviour of the lift cable at the atomic level. A naive mechanical model of the dependency may need to be modified if the possibility of thermal effects is admitted, and is no longer appropriate if the lift cable is presumed to become slack or to break. Nor is dependency necessarily associated with synchronisation of change; it has more to do with what we informally understand by causation, as when a doctor declares that a living person has been fatally wounded. This leads us to view dependency as framing what consequences of an action are an inevitable effect of the action, and cannot be allayed by the intervention of any other agent. There is also an important distinction between the notion of dependency that is often invoked in analysing concurrency in traditional procedural programs, which broadly relates to how the current value of a variable depends on previous assignments to other variables, and the indivisible characteristic of the dependency that in general binds many changes to one atomic change in EM.

There is no explicit reference to dependency in James's account of pure experience. However, amongst the conjunctive relations he lists "relations of activity, tying terms into series involving change, tendency, resistance and the causal order generally" (1996: 44–45). One of the most significant features of EM is that it provides an alternative to the classical procedural model of state, where the values of variables are treated as discrete and independent: the dependencies in a script are not assertions about values alone, but embody expectations about responses to interaction that are themselves a part of state. To the extent that these expectations are relations between observables that are directly experienced, it seems appropriate to classify dependencies as conjunctive relations of activity. A plausible analogy likens the disjoint terms of traditional empiricism to the discrete variables of procedural programming, and the conjunctive relations of Radical Empiricism to the dependencies in modelling with definitive scripts. The elaborate programming mechanisms that are required to maintain dependencies between variables in a procedural representation of state echo James's comment about the complex ways in which rationalists remedy disjunction.

James's perspective on the relationships between transitions in experience is helpful in clarifying the interpretations of interactions and dependencies in EM models. In the EM model depicted in Figure 1, for instance, the lift is conceived as 'moving from one floor to the next' though there is no explicit attempt to model continuous lift motion. Within this naive lift model, the location of the lift user is thought of as defined by 'in the lift', 'on a particular floor' or 'not in the vicinity of the lift'. This 'user location' cannot be identified with an actual physical location, but is nonetheless deemed to be a meaningful observable. In this context, just exactly where the lift user is standing in — or in the vicinity of — the lift, or what posture they adopt, is not significant. From this, it may appear that the dependency by which "the position of the user is determined by that of the lift" is to be construed in a different way from the dependency that links the position of the lift car to that of the end of the supporting cable. The correlation between the movement of the end of the cable and that of the lift car is far more exact, and is easily interpreted with reference to atomic 'infinitesimal' change. To conceive the movement of the lift between one floor and the next (as modelled by `_liftfloor++;`) as an atomic change is to presume that no matter what the lift user does within the lift, they will be moved with the lift from one floor to the next. In the context, the relationship between the continuous motion of the lift and the discrete movement from floor to floor can be interpreted as illustrating what James identifies as substitution, whereby "an experience that knows another can figure as its representative, not in any quasi-miraculous 'epistemological' sense, but in the definite practical sense of being its substitute in various operations, sometimes physical and sometimes mental, which lead us to its associates and results" (James 1996: 61). The natural way in which these two experiences of a lift can co-exist subject to appropriate assumptions about the modeller's motivation is illustrated in the extension of the EM model mentioned above (cf. `liftBeynon2003`). For this purpose, we need only introduce a new observable `liftfloorheight` that (say) can assume integer values from `1` to `5*N` (where `N=10` for instance) corresponding to lift positions at or between floors, and define `_liftfloor` to be the integer part of `liftfloor-height/N`. Such a model is appropriate for many practical purposes, but would not serve for a murder mystery in which we might well be asked to imagine that a person ascending in the lift from floor 3 fails to arrive at floor 4, or meet the need the engineer may have to consider non-integral positions for `_liftfloor` for a lift that is malfunctioning.

As the above discussions indicate, the character of an EM model is moulded by the way that the modeller chooses to interact with it, and how this interaction reflects the modeller's evolving presumptions about the agency to be taken into consideration. In much the same way that each visit to Coventry serves both to fulfil familiar expectations and functions and to introduce what is changed or was previously unknown, interaction with an EM model involves both creation and use. In contrast, the formal specification of computer models requires a commitment to modes of agency and interpretation (i.e., in 'creating' the model) that cannot be reappraised in the subsequent course of interaction with the model (i.e., in 'using' the model). This accounts for a fundamental difference in orientation between EM and traditional computer-based modelling. In the former context, our agenda is 'finding a good construal', for which the key question is 'given that this is what we're interested in achieving, what assumptions about agency in the world is it necessary and appropriate to make?'. In the latter context, our agenda is 'optimising our use of known resources for familiar purposes', for which the key question is 'given that this is the agency in the world, how do we best exploit it?'. The premise for the former agenda is ignorance of the world, and for the latter, knowledge of the world.

From a philosophical perspective, RE can be viewed as endorsing this shift in engineering priorities that EM promotes. In "The Experience of Activity", James (1996) remarks:

> … the healthy thing for philosophy is to leave off grubbing underground for what effects effectuation, or what makes actions act, and to try to solve the concrete questions of where effectuation in this world is located, of which things are the true causal agents there, and of what the more remote effects consist (James 1996: 185–186).

The thrust of agent-oriented analysis in EM is arguably well-aligned to James's recommended agenda in its concern for identifying agency, attributing state-change to agents and interpreting agent interaction in global state-based terms. In contrast to the mainstream traditions of research on agent-oriented modelling and programming, EM favours a concept of 'agency' that is more than any circumscribed preconceived rationalised interaction, and is oriented towards a pragmatic dynamic shaping of construals. This emphasis is consistent with James's recommendation that, in examining 'the real facts of activity', and arbitrating between whether our actions are programmed by a higher authority, are an expression of free will, or emerge from the corporate behaviour of more primitive agents, we should evaluate our responses to the question "Whose is the real activity?" by asking "What will be the actual results?" (James 1996: 178–179). In principle, EM provides a practical framework within which to tackle this agenda, supplying environments in which to explore 'possible construals' and to situate the negotiation of meaning.

As the discussion of the possible extensions of the EM lift model illustrates, the virtue of such an environment is that it is a source of experience that is rich to the point of incoherence. As James (1996: 132–133) observes:

> Experiences come on an enormous scale, and if we take them all together, they come in a chaos of incommensurable relations that we can not straighten out. We have to abstract different groups of them, and handle these separately if we are to talk of them at all.

In the EM model of the lift, we can accommodate the possibility that the lift scenario described in Box 1 occurred not because — as I hypothesised — there was a user Z, but simply because the lift control was faulty. We can dramatise I's predicament in deciding whether to get out of the lift on floor 4 without needing to resolve the logical inconsistencies in his perception and pursuing these to their contradictory conclusions. In contrast, conventional programming, like traditional empiricism, has no satisfactory way to handle the incompleteness of knowledge. Without such means, it cannot do justice to James's conception of ourselves as 'virtual knowers', or his observation that "To continue thinking unchallenged is, ninety nine times out of a hundred, our practical substitute for knowing in the completed sense" (1996: 69).

## 3. Conclusion

The perspective on the nature of knowing that RE and EM endorse has direct practical relevance for current trends in knowledge management (cf. Beynon, Rasmequan, and Russ 2000). To date, the successful application of computers in management has relied to a large extent on exploiting what can be objectified and expressed in formal notations (as in a relational database, or an expert system). As we seek to make yet more sophisticated use of computer technology, and as this technology itself potentially embraces broader aspects of the total business experience, so the limitations of widely accepted philosophies of information science are being exposed. The problems we face are epitomised by the difficulties of negotiating ontologies and standardising formal representations and procedures for communication and re-use. This paper attributes these problems to the barriers that traditional philosophical frameworks for cognition and computation place between words and concepts and the experience that informs them.

The fundamental role that experience plays in informing concepts, as emphasised by both RE and EM, is patent in the everyday situations within which knowledge management has to function. Consider the experience that leads us as we grow up to identify one and the same entity as "a person", "a man", "a doctor", and "a paediatrician", or to appreciate the distinction between 'learning to speak French' and 'being French'. In attempting to capture such concepts in formal ontologies without reference to experience of artefacts, it is arguably impossible to do justice to the role of tacit knowledge, and to reflect the subtle nuances concerning the perceptibility, reproducibility and stability of the underpinning experience. As we aspire to provide computer support for 'experience management', and accommodate such broad perspectives on knowledge such as 'mimetics' affords, there is ever more need to take explicit account of personal experience and to understand this in relation to our interaction with the natural world and with other people. RE in conjunction with EM potentially offers a philosophical and practical framework within which to give proper prominence both to direct experience and the personal stream of thought, and to the distinctions between knowledge as 'socially accepted' and knowledge as 'experientially validated'.

For the sceptical reader, a major intellectual objection to engaging with the thesis of this paper is that it represents RE and EM as essential alternative fundamental philosophical and computational perspectives. To acknowledge that concepts such as 'theory', 'reasoning' and 'reality' have their significant place within this perspective is not enough to deflect such scepticism. In this context, a key issue is that we have become inured to interpreting our interactions with computing technology in a narrow sense that we would (arguably) not entertain as appropriate in relation to other experiences, such as playing or listening to a musical instrument. Indeed, the influence of computational theory on cognitive science has been such that there is a tendency towards construing all interaction as a form of computation. James's characterisation of the nature of knowing is of crucial significance in this connection: it identifies the relationship between one experience and another not as rationally apprehended and explicable with reference to preconceived criteria for similarity, but as itself given in experience. For the sceptic, a useful first step towards appreciating this view of what it is to know is to distinguish the EM model of a lift from an orthodox simplified formal model of a lift with a prescribed and preconceived interpretation and functionality. In the longer term, in the author's opinion, the possibility of wider acceptance of RE and EM as a new framework for knowledge management does not rely upon such intellectual assent: it potentially offers such benefits in terms of the quality of the results and experience it can offer that its practical application will be justification in itself.

## Note

* I am indebted to all the participants in the EM project for their various contributions and in particular to Cheryl Sidebotham for the basic EM lift model. I am especially indebted to Steve Russ for his role in promoting my interest in the philosophical aspects of EM and its application to issues of knowledge management, and for presenting an early version of this paper at the Workshop on Philosophy and Knowledge Management at Lucerne in April 2003. I also wish to thank Bertin Klein and Hans-Joachim Petsche for their help and encouragement.

## References

Beynon, W.M. 1997. "Empirical modelling for educational technology". In *Proc. Cognitive Technology* 1997, University of Aizu, Japan, IEEE, 54–68.

Beynon, W.M. 1999. "Empirical modelling and the foundations of AI". In C.L. Nehaniv (ed), *Computation for Metaphors, Analogy and Agents*. Berlin: Springer, 322–364.

Beynon, W.M., Ch'en, Y.-C., Hseu H.-W., Maad, S., Rasmequan, S., Roe, C., Rungrattanaubol, J., Russ, S.B., Ward, A.T., and Wong, K.T.A. 2001. "The computer as instrument". In M. Beynon, C.L. Nehaniv, and K. Dautenhahn (eds), *Proc. Cognitive Technology: Instruments of Mind*. Berlin: Springer, 476–489.

Beynon, W.M., Rasmequan, S., and Russ, S.B. 2000. "A new paradigm for computer-based decision support". *Decision Support Systems* 33(2): 127–142

Beynon, W.M., Rungrattanaubol, J., and Sinclair, J. 2000. "Formal specification from an observation-oriented perspective". *Journal of Universal Computer Science* 6(4): 407–421.

Beynon, W.M. and Russ, S.B. 1991. "The development and use of variables in mathematics and computer science". In J.H. Johnson and M. Loomes (eds), *The Mathematical Revolution Inspired by Computing*. Oxford: Oxford University Press, 85–95.
Beynon, W.M. and Sun, P.-H. 1999. "Computer-mediated communication: A distributed empirical modelling perspective". In *Proc. Cognitive Technology* 1999, San Francisco, 115–132.
Bird, G. 1986. *William James*. London: Routledge and Kegan Paul.
Cantwell-Smith, B. 2002. "The foundations of computing". In M. Scheutz (ed), *Computationalism: New Directions*. Cambridge, MA: The MIT Press.
Ginsberg, M.L. and Smith, D.E. 1988. "Reasoning about action I: A possible worlds approach". *Artificial Intelligence* 35: 165–195
Gooding, D. 1990. *Experiment and the Making of Meaning: Human Agency in Scientific Observation*. Dordrecht: Kluwer.
Jacobson, I., Christerson, M., Jonsson, P., and Overgaard, G. 1992. *Object-oriented Software Engineering: A Use-Case Driven Approach*. New York: Addison-Wesley.
James, W. 1996 [1912]. *Essays in Radical Empiricism*. New York: Dover.
Levi-Strauss, C. 1966. *The Savage Mind*. Chicago: The University of Chicago Press.
Rungrattanaubol, J. 2002. *A Treatise on Modelling with Definitive Scripts*. PhD Thesis, Computer Science Department, University of Warwick.
Turner, M. 1996. *The Literary Mind*. Oxford: Oxford University Press.
The EM website: http://www.dcs.warwick.ac.uk/modelling/
The EM archive: http://empublic.dcs.warwick.ac.uk/projects/

# Index

In the series *Benjamins Current Topics (BCT)* the following titles have been published thus far or are scheduled for publication:

12 **DROR, Itiel E. (ed.):** Cognitive Technologies and the Pragmatics of Cognition. 2007. xi, 185 pp.

11 **PAYNE, Thomas E. and David J. WEBER (eds.):** Perspectives on Grammar Writing. vii, 213 pp. + index. *Expected September 2007*

10 **LIEBAL, Katja, Cornelia MÜLLER and Simone PIKA (eds.):** Gestural Communication in Nonhuman and Human Primates. xiv, 275 + index. *Expected September 2007*

9 **PÖCHHACKER, Franz and Miriam SHLESINGER (eds.):** Healthcare Interpreting. Discourse and Interaction. 2007. viii, 155 pp.

8 **TEUBERT, Wolfgang (ed.):** Text Corpora and Multilingual Lexicography. 2007. x, 162 pp.

7 **PENKE, Martina and Anette ROSENBACH (eds.):** What Counts as Evidence in Linguistics. The case of innateness. 2007. x, 297 pp.

6 **BAMBERG, Michael (ed.):** Narrative – State of the Art. 2007. vi, 271 pp.

5 **ANTHONISSEN, Christine and Jan BLOMMAERT (eds.):** Discourse and Human Rights Violations. 2007. x, 142 pp.

4 **HAUF, Petra and Friedrich FÖRSTERLING (eds.):** Making Minds. The shaping of human minds through social context. 2007. ix, 275 pp.

3 **CHOULIARAKI, Lilie (ed.):** The Soft Power of War. 2007. x, 148 pp.

2 **IBEKWE-SANJUAN, Fidelia, Anne CONDAMINES and M. Teresa CABRÉ CASTELLVÍ (eds.):** Application-Driven Terminology Engineering. 2007. vii, 203 pp.

1 **NEVALAINEN, Terttu and Sanna-Kaisa TANSKANEN (eds.):** Letter Writing. 2007. viii, 160 pp.